PROJECT AIR FORCE

T0127930

An Integrated Survey System for Addressing Abuse and Misconduct Toward Air Force Trainees During Basic Military Training

Kirsten M. Keller, Laura L. Miller, Sean Robson, Coreen Farris, Brian D. Stucky, Marian Oshiro, Sarah O. Meadows

Prepared for the United States Air Force

Approved for public release; distribution unlimited

For more information on this publication, visit www.rand.org/t/RR964

Library of Congress Cataloging-in-Publication Data is available for this publication.

ISBN: 978-0-8330-8893-2

Published by the RAND Corporation, Santa Monica, Calif.

© Copyright 2015 RAND Corporation

RAND® is a registered trademark.

Support RAND

Make a tax-deductible charitable contribution at

www.rand.org/giving/contribute

www.rand.org

Preface

In response to several high-profile incidents of sexual misconduct by military training instructors during U.S. Air Force Basic Military Training (BMT), in 2012 the Air Force's Air Education and Training Command asked RAND Project AIR FORCE to help develop an integrated survey system to help address abuse and misconduct toward trainees in the BMT environment. This report provides an overview of the survey system, including recommendations for administration, analysis, and the reporting of results. The report also outlines how this survey system fits into the broader BMT leadership feedback system and identifies additional gaps or areas for improvement to better track and monitor actual instances of and the potential for abuse and misconduct.

The research reported here was commissioned by the commander of the Air Force's Air Education and Training Command (AETC/CC) and was conducted within the Manpower, Personnel, and Training Program of RAND Project AIR FORCE.

RAND Project AIR FORCE

RAND Project AIR FORCE (PAF), a division of the RAND Corporation, is the U.S. Air Force's federally funded research and development center for studies and analyses. PAF provides the Air Force with independent analyses of policy alternatives affecting the development, employment, combat readiness, and support of current and future air, space, and cyber forces. Research is conducted in four programs: Force Modernization and Employment; Manpower, Personnel, and Training; Resource Management; and Strategy and Doctrine. The research reported here was prepared under contract FA7014-06-C-0001.

Additional information about PAF is available on our website:
http://www.rand.org/paf/

This report documents work originally shared with the U.S. Air Force incrementally during the fall of 2013 and winter of 2014 in the form of briefings, survey instruments, and survey data templates. The draft report, issued on August 2014, was scrutinized by formal peer reviewers and U.S. Air Force subject-matter experts.

Table of Contents

Figures and Tables

Figure

Tables

Summary

In response to several high-profile incidents of sexual misconduct by military training instructors (MTIs) during U.S. Air Force Basic Military Training (BMT), in 2012 the Air Force's Air Education and Training Command (AETC) asked RAND Project AIR FORCE to develop an integrated survey system to address abuse and misconduct toward trainees in the BMT environment. Each year, about 35,000 new Air Force recruits attend BMT at Lackland Air Force Base, Texas, where they spend about eight weeks before attaining the status of airman. They then move to different locations throughout the country for initial skills training (IST)—where they will gain the basic knowledge and skills required for their respective Air Force job specialties— and, later, for their first duty station assignments.

In response to the AETC request, RAND developed two complementary surveys—one for trainees and one for MTIs. RAND also recommended guidelines for administering the surveys, analyzing the data, and reporting results. The AETC commander established two primary purposes for the surveys: (1) to help detect incidents of abuse and misconduct in the training environment, and (2) to provide data to help leaders understand what actions to take to reduce abuse and misconduct. In consultation with AETC headquarters and BMT leadership, RAND identified the following five core domains of abuse and misconduct to assess through the surveys:

- trainee bullying
- maltreatment and maltraining by MTIs
- unprofessional relationships with MTIs
- sexual harassment from anyone
- unwanted sexual experiences committed by anyone.

To develop the survey system, RAND undertook several phases of information gathering and tested the draft survey items and administration procedures. RAND researchers built on multiple investigations conducted just prior to the start of this research, including trend reports on MTI staffing and reported incidents, focus groups, and a survey of every trainee in BMT in the summer of 2012. We also reviewed results of previous end-of-course trainee surveys and MTI quality-of-life surveys. To the extent possible, the research team drew on established measures in the scientific literature with strong psychometric properties (i.e., the scale consistently and appropriately measures the construct it intends to measure) for inclusion into the surveys. In some cases, though, existing measures are not well suited to the BMT context, or the constructs being measured are BMT specific and there are no existing measures (e.g., maltreatment and maltraining, unprofessional relationships). In these instances, we developed new items and scales specifically for these surveys, which were put through multiple reviews by subject matter experts and other key AETC stakeholders. We then conducted an initial test of both the trainee and MTI survey items by administering the draft surveys to trainees and MTIs currently at BMT. This allowed us to make decisions about final content based on our analyses of the test survey data (e.g., item and scale performance), written comments and suggestions of participants, considerations of how best to package and report results, and final feedback from AETC and BMT leadership.

BMT Trainee Survey

The BMT trainee survey provides a framework for assessing the prevalence and reporting of abuse and misconduct associated with trainee bullying, maltreatment and maltraining by MTIs, unprofessional relationships with MTIs, and sexual harassment and unwanted sexual experiences perpetrated by anyone at BMT. To help guard against trainees being identified with their survey responses, the survey includes only one background question—the trainee's gender. This is the single most important background variable for understanding the survey results due to potential differences in male and female experiences with the types of abuse and misconduct assessed on the surveys. Because of the limited number of women in BMT, including additional background questions, such as training squadron or race/ethnicity, would increase the risk that a female respondent could be identified or would be concerned about being identified.

All of the abuse and misconduct sections of the survey focus on behavior directed at trainees. The bullying section is designed to focus specifically on trainees' abuse of other trainees. The maltreatment and maltraining section focuses on abuse of trainees from MTIs, including the use of any training practice not designed to meet a course training objective (e.g., excessive physical training). Survey questions about unprofessional relationships between MTIs and trainees address inappropriate exchanges of money, inappropriate social contacts, and other relationship-policy violations (e.g., drinking alcohol with a trainee). Finally, the survey includes questions on sexual harassment and unwanted sexual experiences from anyone at BMT (e.g., trainees, MTIs), with a series follow-up questions for trainees who indicate that they have been harassed or had an unwanted sexual experience.

Within each abuse and misconduct section of the survey, there are also items that ask trainees about their experience reporting or telling others of that type of abuse and misconduct. This is then followed by questions assessing trainees' perceptions of the squadron climate in terms of the extent to which squadron leaders enforce abuse and misconduct policies. The survey concludes with items about BMT leaders, support personnel, and feedback and support systems. A final question asks trainees whether they felt comfortable being open and honest on the survey, which can be an indication of the level of trust at BMT and the success of the survey administration procedures. It also serves as a way to screen responses.

MTI Survey

A complement to the trainee survey, the MTI survey assesses the extent to which MTIs are aware of abuse and misconduct directed toward trainees. It also includes questions to assess MTIs' perceptions of the squadron climate on abuse and misconduct, whether MTIs as a whole are willing to report such incidents, and the clarity of abuse and misconduct policies.

The MTI survey also examines MTI quality of life, including job attitudes, perceptions of the work environment, job stressors, and professional development. Prior to this RAND project, AETC had been administering an MTI Quality of Life Survey (QOLS) once a year. In order to minimize the risk of survey burnout among MTIs, AETC requested that RAND review the QOLS and determine whether the constructs it measured could be integrated into RAND's survey. We found that in many ways the QOLS survey was already aligned with the purposes of the new abuse and misconduct survey, so we were able to integrate QOL constructs in the survey we developed. An MTI's job performance is defined not only by the ability to prepare trainees to become airmen but also the extent to which maltraining, maltreatment, unprofessional relationships, and harassment of trainees are avoided. Moreover, MTIs are responsible for ensuring that their fellow MTIs also comply with these policies and report any known or suspected violations. Whether MTIs meet these expectations could be influenced to some extent by the attitudes they have toward the Air Force, BMT, trainees, and their jobs as MTIs.

Survey Participation and Administration

We recommend the surveys be administered to both trainees and MTIs by computer to permit branching on survey questions. Computerized surveys can also help minimize data recording errors and the time needed to complete and analyze the survey.

We recommend that the trainee survey be administered to trainees through individual computer stations within a BMT classroom, although participation should be voluntary. Because some abuse and misconduct behaviors may be rare, surveying all trainees, rather than just a sample, will provide better prevalence estimates. It may also help prevent offenders from targeting groups they know will not be surveyed in an effort to reduce the risk of detection. We recommend that all trainees be required to remain in the survey room for the entire survey session. Permitting trainees to leave as they complete their surveys would provide incentive for them to decline to participate or rush through the survey, and could suggest to peers that trainees who take more time are indicating abuse and misconduct and thus are receiving more of the follow-up questions. We also recommend that trainees take the survey as close as possible to the end of BMT and as an exit survey for trainees who fail to complete BMT.

Assuming that no other routine MTI surveys are introduced, we recommend conducting the MTI survey every six months to a year, with more-frequent surveys following spikes in abuse and misconduct incidents, major personnel turnover, and changes to policies or programs that could have an intentional or unintentional impact on the elements measured by the survey. To ensure that participants have sufficient knowledge to respond to the survey, we also recommend that the MTI survey be given only to MTIs who have served at BMT for at least one month.

To promote honest responses and protect survey respondents, especially victims, the surveys should be anonymous or confidential to the greatest extent possible. The point is to provide both actual protection and the perception of protection among potential respondents. For example, the instruments we propose greatly limit demographic questions. We also recommend that respondents be able to participate without having to enter identifying information (including using a common access card for computer access), and without MTIs having to use their own personal computers or a computer assigned to their workstations. We recommend allowing only participants and a civilian survey administrator in the survey room, so no authority figures are present. During the test version of the survey, AETC also outfitted the computer screens with privacy protectors to reduce the risk that neighboring survey participants could see one another's responses. Finally, the survey data should be protected through password protection and encryption to reduce unauthorized access, and the results should never be reported in a way that might permit the identities of individual participants to be deduced. For example, there may be only one trainee who formally reported a sexual assault in a single quarter, so that trainee's responses should not be presented individually in briefings or reports.

Reporting Results and Taking Action

Assessment is only the first step toward improvement. A fully effective feedback system follows assessment with (1) analyses and tracking trends over time, (2) triangulation with other relevant data sources and follow-up data collection to better understand the results, (3) a systematic process for reporting results to senior Air Force leaders and other key stakeholders, (4) prioritization of problem areas and setting goals for improvement, and (5) taking action to implement new policies and improvement plans.

We strongly recommend that a qualified analyst with a background in the social sciences and statistics conduct the analyses and interpret the results. This will provide AETC with in-house analysis expertise to clarify whether changes reflect expected fluctuations or represent trends of concern or improvement. Data interpretation and additional insights into abuse and misconduct problems and solutions can also be gained by triangulating the survey data with other data sources, including official channels for reporting abuse and misconduct or as-needed focus groups and follow-up interviews. Given the different types of information and frequency with which they are to be collected, we recommend reporting trainee survey results on a quarterly basis and reporting MTI survey results every six months or yearly (following each MTI survey). Reports should be tailored to each leadership level and relevant groups of stakeholders. To prioritize problem areas, leadership should engage MTIs and other stakeholders to develop criteria. Leaders should also engage stakeholders when acting on the results to ensure that changes are being accepted, followed, and work as intended.

Conclusion and Additional Recommendations for BMT

Just as the survey system is designed to augment rather than substitute for direct interaction and monitoring by leadership and support professionals, there are other supplemental actions the AETC leadership could take to improve the training environment and prevent abuse and misconduct. We offer several additional observations for consideration:

- **Routinely monitor security camera footage.** AETC has installed additional security cameras throughout the BMT area, but footage is only reviewed if a complaint has been registered. Those recordings could be monitored for abuse, misconduct, and other prohibited behavior, and thus provide a way to detect incidents that might not otherwise be reported.

- **Evaluate the training that prepares trainees to identify and report abuse and misconduct.** Trainee education and training related to abuse and misconduct have been substantially revised since 2012. This training is the process through which BMT teaches trainees about their expected roles and responsibilities within the leadership feedback system. We recommend evaluating that training every few years to assess whether trainees comprehend and apply the material as AETC intends. In other words, are they able to sufficiently identify what behavior they should be reporting and the various reporting channels they can use to alert leadership to problems or to seek help? Trainees should have a good understanding of which MTI training techniques are inappropriate and should be reported, and of what constitutes sexual harassment and assault so they can conduct themselves accordingly and respond appropriately if they witness or experience this behavior.

- **Follow up with victims and witnesses who have filed reports of sexual assault.** Feedback on the experiences of trainees who have reported sexual assault or other serious complaints of abuse and misconduct can help AETC leadership identify negative experiences that violate standards or policy and could deter other victims and witnesses from coming forward. The Air Force should explore feedback mechanisms for understanding victims' and witnesses' experiences with and perceptions of the reporting process, the judicial system, the victim care/advocacy system, and unit leadership reactions to their report. Someone outside these systems, such as a sexual assault response coordinator (SARC) or victim advocate from another major command, could be asked to reach out to trainees who have filed reports of incidents and ask that they volunteer to discuss their experiences. Note, we are not advocating trying to use the BMT survey to identify and follow up with trainees who have indicated having an unwanted sexual experience. As trainees move into IST or their first duty station, the assessment could also address the continuity of support and care for victims across these transitions. Feedback efforts will have to be developed with great care to ensure that they are not intrusive to victims, and that access to information about the identities of those who filed restricted reports continues to be severely restricted. SARCs already often follow up with victims as a part of care coordination: a new feedback system could at a minimum request aggregate, standardized updates from SARCs similar to the aggregate, standardized reports that SARCs prepare about initial reports of sexual assault.

- **Create an online central repository accessible by key leadership and support professionals.** We recommend building an online central repository for sharing the aggregate data from the surveys and the additional sources of feedback we identified as indicators of abuse and misconduct. Access to this repository could be limited only to those with a legitimate use for it. As an archive, this repository would also protect against the loss of historical trend data. This repository would better enable leaders within BMT to connect the dots across these sources of information and to identify vulnerabilities in the system.

In conclusion, AETC has made great strides toward increasing its monitoring of the BMT environment and implementing reforms to improve its ability to dissuade, deter, detect, and hold accountable those responsible for abuse and misconduct. The survey system developed by RAND provides a way for trainees and MTIs to report abuse and misconduct toward trainees confidentially and without the fear of embarrassment or reprisal. It makes a unique contribution to the leadership feedback system that grows as data are accumulated. By institutionalizing this survey, AETC has ensured that leaders will be alerted in a timely manner to abuse and misconduct long after the subject has disappeared from the headlines.

Acknowledgments

The development of this survey system, which was aligned to fit Air Education and Training Command (AETC) priorities, needs, and capabilities, was possible only through a great deal of coordination with key staff at AETC.

Air Force General Edward A. Rice Jr. (now retired) was the AETC commander and project sponsor during the course of this research. General Rice adopted the recommendation to develop an anonymous survey system from the 2012 commander-directed investigation of basic military training led by Major General Margaret H. Woodward (now retired). General Rice established the overarching goals for the survey system and made the local resources available to develop and test it.

Major General Timothy M. Zadalis served as a project monitor from AETC for most of the project, before he moved to another assignment. His thoughts and guidance helped shape the direction of this project.

Colonel John David W. Willis, our primary project monitor, was actively engaged across the entire project and contributed insightful and practical feedback on the approach and instruments. He also went to great lengths to facilitate on-site administration of the test version of the survey that did not require participants to use identifying information to access the surveys.

We are also grateful for the involvement of Dr. Laura Munro, who shared valuable information and perspectives about Air Force Basic Military Training (BMT) and also facilitated administration of the test surveys and provided feedback on elements of the survey system. Additionally, Bob Wilson made important contributions through sharing his expertise of previous surveys of trainees and trainers at BMT that he has conducted and analyzed.

Jim Steele and Kevin McGaughey assisted with preparations and provided on-site support during the administration of the test versions of the survey. Colonel Jeanne Hardrath and Senior Master Sergeant Kevin Richards facilitated participation of a group of recent BMT graduates in our test survey. James Smith originated the concept, and Toni Webster, David Gorham, and the team at AETC/A6 contributed to the development of the computer solution used to administer the surveys. William Hall, the BMT sexual assault response coordinator, arranged for a sexual assault response coordinator or victim advocate to be in an adjacent room during all of our survey sessions in case any trainees became distressed by the survey subject matter and needed help (fortunately, none did).

A number of key staff in AETC contributed critical information and feedback on early survey drafts to ensure that the content and the language fit the BMT environment, leadership concerns, and Air Force definitions and policy. In addition to those already named, these include but are not limited to Major General Leonard Patrick, Chief Master Sergeant Gerardo Tapia, Colonel Lane Benefield, Christine Burnett, Bonnie Molina, Colonel Robert Miller, Colonel Roosevelt

Allen Jr., Colonel Deborah Liddick, Lieutenant Colonel Timothy Owens, Roger Corbin, Cindy Luster, and Jose Caussade. Additionally, Manuel Sancillo's questions and suggestions through the initial implementation of the surveys helped us refine and fill out our recommendations for survey programming, content, and reporting. Feedback from AETC key experts on an earlier draft of this report helped us to clarify several points and to ensure accuracy of factual information about AETC. Captain Shawna Parker and Captain Vincenza Grossman provided logistical support for installation visits.

The RAND Project AIR FORCE (PAF) manager Carl Rhodes provided valuable feedback throughout the course of this project. Air Force Lieutenant Colonel Todd Osgood supported the project while he was an Air Force fellow at RAND through contributions such as feedback on draft products and research assistance. Lara Schmidt led an intensive early quality assurance review of the proposed survey administration mode and the trainee survey instrument. Our peer reviewers in this process, Claudia Bayliff, attorney at law, and RAND colleagues Bonnie Ghosh-Dastidar and Meg Harrell, offered rigorous and constructive reviews that led to improvements reflected in this report. We also benefited from the peer review of an earlier version of draft report that was provided by Bayliff and RAND's Kimberly Hepner and Lane Burgette, as well as feedback from the PAF manager Ray Conley. The MMIC^{TM} (Multimode Interviewing Capability) programmers Julie Newell and Adrian Montero assisted with programming the test survey instrument. Adrian also helped prepare the virtual computing environment at Lackland Air Force Base so that administration could run smoothly and the data could be appropriately safeguarded. Dionne Barnes assisted with administration of the test version of the survey.

RAND's institutional review board, the Human Subjects Protection Committee, also provided a very constructive review of our proposed survey methods. Jessica Candia and Kathy Sehhat in the Research Oversight and Compliance Division of the Office of the Air Force Surgeon General expedited their review of this project so that we could meet the time frame in which General Rice needed the survey system components.

We are also thankful for editorial support. Kristin Leuschner edited the draft and final versions of the trainee and MTI surveys and helped to create the trainee survey flow chart in Chapter Two. Melissa Bauman and Kate Giglio helped by editing the chapters in this report, and Cliff Grammich helped write the summary. Jonathan Martens assisted with preparation of the reference list and earlier in the project helped compile AETC feedback used in developing some of the draft survey items. Finally, the writing was improved by Rebecca Fowler's editing of the completed final manuscript, and Mary Wrazen prepared the final version of the trainee survey flow chart.

Abbreviations

AETC	Air Education and Training Command
BEAST	Basic Expeditionary Airman Skills Training
BMT	Basic Military Training
CAC	common access card
CDI	commander-directed investigation
CFA	confirmatory factor analysis
CFI	comparative fit index
DMDC	Defense Manpower Data Center
DoD	Department of Defense
EFA	exploratory factor analysis
FT	flying training
ISH	Inventory of Sexual Harassment
IST	initial skills training
MTI	military training instructor
NCO	noncommissioned officer
OSI	Office of Special Investigations
PT	physical training
QOLS	Quality of Life Survey
RAINN	Rape, Abuse and Incest National Network
RETOC	Recruiting, Education and Training Oversight Council
RMSEA	root mean square error of approximation
SARC	sexual assault response coordinator
SEQ	Sexual Experiences Questionnaire
SF	Security Forces
SHI	Sexual Harassment Inventory
SME	subject matter expert

TLI	Tucker Lewis index
TT	Technical Training
UCMJ	Uniform Code of Military Justice
WS/MS	Warrior Skills and Military Studies

1. Introduction

Every year, roughly 35,000 new U.S. Air Force recruits attend Basic Military Training (BMT) at Lackland Air Force Base near San Antonio, Texas. BMT is the Air Force's entry-level training (i.e., basic training or boot camp) that all enlisted recruits must pass through and graduate before they are considered airmen.[1] It is a brief, intense period of knowledge acquisition and skill building that is intended to provide foundational training and socialization into Air Force culture. At the time of this study, BMT was approximately eight weeks long. After completing and graduating from BMT, trainees move on to locations throughout the country for initial skills training (IST), where they will gain the basic knowledge and skills required for their respective Air Force job specialties. IST is composed of flying training (FT) and technical training (TT) courses. Finally, new airmen are assigned to their first duty stations.

Enlisted noncommissioned officers (NCOs) known as military training instructors (MTIs) provide the majority of new recruits' training in BMT. MTIs are responsible for motivating, disciplining, and instructing trainees as they progress through BMT. BMT is what sociologists call a *total institution*: members live segregated from society in an environment with routines that are tightly scheduled and lives highly controlled by authority figures (Goffman, 1961). MTIs control almost every aspect of trainees' lives in BMT, including when they are allowed to eat, sleep, and talk. Trainees' privacy and privileges are severely restricted: Some freedoms are gradually earned over the course of BMT, and some—such as leaving Lackland Air Force Base or having outside visitors—are not usually permitted prior to graduation. MTIs hold considerable power over trainees and help determine whether a trainee is fit to graduate and join the Air Force. As a result, this organizational structure and power imbalance can present opportunities for MTIs to abuse their authority.

In 2012, Air Education and Training Command (AETC) found multiple MTIs guilty of sexual assault and unprofessional relationships with trainees. Several AETC-directed reviews of the circumstances led to recommendations for extensive reform in the training environment. As part of these reforms, the commander of AETC asked RAND Project AIR FORCE to assist AETC in fostering a BMT environment free of abuse and misconduct through the development of a standardized survey system designed to improve the monitoring of behavior, attitudes, and the overall Air Force approach to preventing abuse and misconduct by instructors.

---

[1] The terms *airman* and *airmen* apply to all officers and enlisted Air Force personnel, men and women, regardless of rank or active-duty, guard, or reserve status. This could potentially be confused with the names of the lowest enlisted ranks in the Air Force, which are airman basic (E1), airman (E2), airman first class (E3), and senior airman (E4), but in this report we use the term only in the general sense.

Objective and Analytical Approach

AETC established two primary purposes for the survey system: help detect abuse and misconduct in the training environment and provide data to help leaders identify ways to reduce abuse and misconduct. To meet these objectives, RAND developed two complementary surveys—one for trainees and one for MTIs. The trainee survey is designed to assess trainees' experiences of abuse and misconduct at BMT and identify any barriers that prevent them from reporting these incidents. The MTI survey is designed to complement the trainee survey and assesses the extent to which MTIs are aware of abuse and misconduct taking place. The MTI survey also examines their attitudes, perceptions of the work environment, and stressors that may influence their ability to prevent and respond to abuse and misconduct.

In consultation with AETC headquarters and BMT leadership, we identified the following five core domains of abuse and misconduct to assess with the surveys:

- trainee bullying
- maltreatment and maltraining by MTIs
- unprofessional relationships with MTIs
- sexual harassment from anyone
- unwanted sexual experiences committed by anyone.

The decision to focus on these five domains was based not only on the misconduct found in 2012 but also on BMT's existing policies governing appropriate interactions between MTIs and trainees and among the trainees themselves. It is important to note that although these domains are not equivalent in severity, some of these abuse and misconduct behaviors may cluster together or lead to one another. For example, detection of less severe behaviors occurring at BMT can provide an opportunity to address the issue quickly before the situation potentially escalates into greater abuse and misconduct (e.g., sexual harassment and the potential for sexual assault).

To develop the surveys, we undertook several phases of information gathering and reviews, and we tested the draft survey items and administration procedures (Appendix A provides a detailed overview of the process). We were able to build on multiple investigations that had been conducted just prior to the start of our research, including trend reports on MTI staffing and reported incidents, focus groups, and a survey of every trainee at BMT in the summer of 2012. We were also able to review the results of previous end-of-course trainee surveys and MTI quality-of-life surveys. Throughout the process, we regularly engaged different levels of leaders from AETC headquarters, BMT, and support providers, who reviewed and provided feedback on in-progress research briefings and draft content. Groups of trainees and MTIs also offered survey draft critiques and suggestions that influenced the wording and content. Our efforts were also informed by reviews of Air Force and Department of Defense (DoD) policies, the Uniform Code of Military Justice (UCMJ), previous reports from special task forces or commissions investigating relevant topics within DoD, and the scientific literature. We also conducted a rigorous RAND-managed peer review of the draft trainee survey and the proposed survey administration plan. Changes that resulted from that review were also applied to the MTI survey. In July 2013, the surveys were then tested with 240 MTIs and 1,042 trainees. Our test surveys contained many more items than we intended to use in the final versions presented in this report as well as many write-in options. This allowed us to make decisions about final content based on our analyses of the test survey data, written comments and suggestions of participants, considerations of how best to package and report results, and final feedback from AETC and BMT leadership.

Organization of the Report

The remaining chapters in this report document the content of the surveys and recommendations for their use. Specifically, Chapters Two and Three describe the content of the trainee and MTI surveys. Chapters Four and Five then follow with recommendations for administration and the reporting of the survey results. Finally, Chapter Six presents our concluding thoughts and additional recommendations for enhancing the overall leadership feedback system at BMT.

The report also includes a number of appendixes. Appendix A provides greater methodological details on the development of the survey content. Appendixes B and C contain the trainee and MTI surveys. Appendix D provides a snapshot of the reporting template developed to help analyze data and track trends. Appendix E provides a sample MTI survey recruitment letter. Appendix F presents an overview of additional data sources at BMT and discusses how these sources, along with the newly developed surveys can help form an integrated feedback system for addressing abuse and misconduct.

2. The BMT Trainee Survey

The BMT trainee survey was developed to help AETC headquarters and BMT leaders detect incidents of abuse and sexual misconduct and to provide data to identify actions needed to prevent and respond to future incidents. This chapter is dedicated to describing the content of the final survey RAND developed, which can be reviewed in full in Appendix B. We also describe the recommended analytic procedures for the survey in this chapter. For the interested reader, we present greater detail on the development of the survey in Appendix A.

BMT Trainee Survey Overview

The overall design of the BMT trainee survey provides a framework for assessing the prevalence and related reporting behaviors of abuse and misconduct in each of the five following core abuse and misconduct domains:

- Section I: trainee bullying
- Section II: maltreatment and maltraining by MTIs
- Section III: unprofessional relationships with MTIs
- Section IV: sexual harassment from anyone at BMT
- Section V: unwanted sexual experiences committed by anyone at BMT.

There is also a sixth section that asks trainees to consider the feedback and support systems available to them at BMT. Based on a test of the survey, the estimated average time to complete the survey is 15–20 minutes. However, some trainees may need more time, depending on how quickly they read and whether their experiences prompt follow-up questions.

Over the course of the first five survey sections, trainees are asked to indicate the frequency with which they experienced different behaviors, whether they were aware of other trainees experiencing the behaviors, and whether they reported or told others of any incidents. Trainees who personally experienced an incident or were aware of an incident had additional questions to answer. This separate series of questions asks them why either they chose not to tell someone at BMT about the behavior or, if they did tell someone, it asks them to provide additional details about their experiences (e.g., how seriously they felt their reports were taken). The survey also includes items to assess individual perceptions of the squadron climate in terms of the extent to which squadron leaders enforce laws and policies related to each of the abuse and misconduct domains. Figure 2.1 presents an overview of the survey flow.

Figure 2.1. Trainee Survey Flow Chart

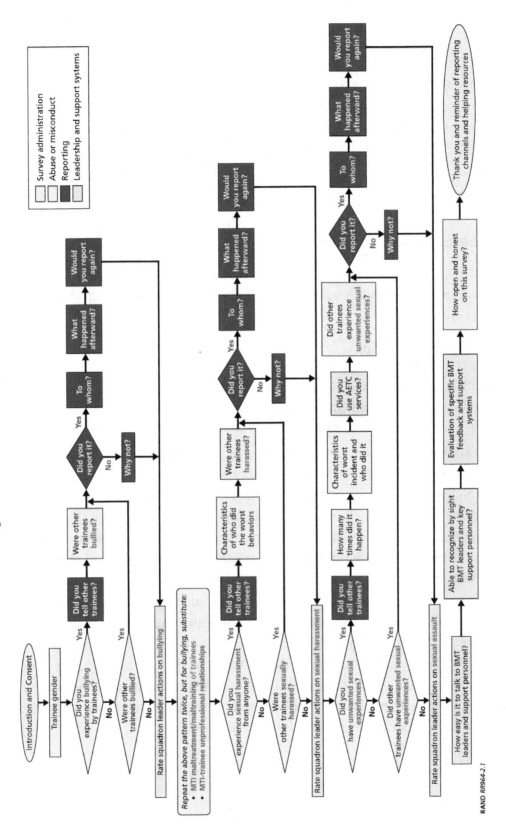

RAND RR964-2.1

6

Descriptions of BMT Trainee Survey Sections

Background: Demographics

To help ensure anonymity, the survey includes only a single background question, which asks for the trainee's gender. This is the single most important background variable to include for understanding the survey results due to potential differences in men's and women's experiences regarding unprofessional relationships, sexual harassment, and unwanted sexual experiences. Because of the limited number of women in some week groups (training cohorts) at BMT, including additional background variables on the survey (e.g., age, race/ethnicity, squadron) would increase the risk of being able to identify an individual trainee and could compromise open and honest participation. For example, although it could be helpful to have information on whether certain racial or ethnic groups tend to be targeted, the relatively small number of female minority members increases the risk of their identities being deduced through their survey responses.[2] The inclusion of the additional background question could also increase the likelihood that these participants would feel uncomfortable being open and honest on the survey.[3]

Sections I–V: Selected Categories of Abuse and Misconduct

Understanding the number of trainees who have experienced different types of abuse and misconduct at BMT can help to inform decisionmakers about how best to design and implement prevention programs and intervention strategies. Furthermore, precise dissemination of resources (i.e., where and to whom they are needed) depends in large part on accurate research to document the prevalence of the behavior in the given population.

To assess the prevalence of behaviors within each abuse and misconduct domain, the first five sections of the survey contain a series of questions asking trainees to indicate the frequency with which they experienced particular types of incidents of abuse or misconduct while at BMT. Frequency is assessed using a five-point response scale, ranging from "never" to "daily,"[4] except

---

[2] As of September 2013, 67 percent of women in the initial enlisted pay grade of E1 were identified as white in Air Force administrative records. To illustrate the challenge of small numbers, only five women were identified as being of Hispanic ethnicity, ten women as Asian, and 285 women as black (the largest minority group). Those women would be spread out across different units (not necessarily evenly distributed) with different graduation weeks, and thus they would be taking the survey different weeks. To view details on these and other racial groups and pay grades, see Culhane, 2014.

[3] We did not collect data on whether the addition of other background questions would affect participants' feelings of anonymity and willingness to answer honestly on the survey. If desired, AETC could add a few questions at the end of the survey about that or conduct survey experiments to examine this hypothesis.

[4] The research team considered using actual numerical counts of incidents as a response option. However, we decided that counts may be difficult for trainees to assess for some of the less severe behaviors that may happen

for the unwanted sexual experiences items, which utilize a "yes or no" response. To assess the extent to which bystanders may be aware of abuse and misconduct taking place, a single follow-up question also asks whether the trainee is aware of other trainees experiencing any of the incidents described in that section.

It is important to note that these survey sections were not designed to describe every conceivable form of abuse and misconduct in these categories. Further, to limit the length of the surveys, we advise against extending the lists unless absolutely necessary to capture a behavior not encompassed by existing items and frequent or severe enough to warrant it. Additionally, the survey does not assess the extent to which trainees label their experiences as "bullying" or "sexual harassment," for example. In the BMT context, the behaviors included on this survey violate policy or law regardless of how the victim might label them or whether the victim agrees with the policy. More specifically, even if a trainee indicated that he or she was not offended by derogatory language being used about gender and did not feel sexually harassed, the use of that language still violates policy and is unwelcome at BMT.

Section I: Trainees Bullying Other Trainees

Much of the abuse and misconduct sections of the trainee survey focus on the extent to which trainees may have experienced abuse from MTIs. However, given the intensity of BMT and required close interactions among trainees, there is also considerable potential for abuse to take place among trainees themselves. Although the BMT trainee code of conduct does not use the term *bullying* specifically, the rules do state that trainees are required to act in a respectful, professional manner at all times, including when interacting with other trainees. This means that bullying-type behaviors are not acceptable.

Definitions of bullying vary somewhat, but in general bullying involves intentional hurtful behavior by one or more individuals that is repeated over time. Bullying is usually characterized by an imbalance of power and can be direct in nature, involving verbal or physical aggression, or indirect, involving social exclusion and manipulation (Solberg and Olweus, 2003). Many bullying behaviors may also capture potential hazing incidents that might occur among trainees, although distinctions between bullying and hazing often rest on the motive for the behavior.

Section I of the survey is designed to assess bullying among trainees. Although a wide variety of scales have been developed to measure bullying, many established scales were not appropriate for the BMT context or were copyrighted for use. Therefore, we instead built on those established scales to develop six items tailored to trainees in BMT. To be consistent with best practices for measuring bullying, the items ask trainees to rate how often they experienced

---

more frequently, such as the use of abusive or offensive language. However, for the unwanted sexual experiences section, the survey does include a follow-up question asking how many times a trainee had an unwanted sexual experience.

specific bullying behaviors from another trainee, including social exclusion and manipulation, theft, threats, and physical violence (five-point response scale ranging from "never" to "daily").

Section II: MTI Maltreatment and Maltraining of Trainees

Maltreatment and *maltraining* are specific terms used by the Air Force to refer to a wide range of behaviors involving MTI abuse of trainees or improper training techniques that may have once been permitted but are not today:

> Maltreatment (physical)—Includes, but is not limited to, poking, hitting, thumping, pushing, grabbing, threats of violence, physical violence, physical intimidation, hazing, or any unnecessary physical contact.

> Maltreatment (verbal)—Any language that degrades, belittles, demeans, or slanders an individual or group based on color, national origin, race, religion, age, ethnic group, gender, or physical stature. Includes, but is not limited to, (1) the use of profanity and any insinuation of immoral, unethical, illegal, or unprofessional conduct; (2) crude, offensive language in rhymes or prose as memory devices (mnemonics); and/or (3) training tools that contain profane words, offensive language, or inappropriate sexual or gender references. Any language that establishes a hostile environment constitutes and promotes sexual harassment, or disrespect to men and/or women.

> Maltraining—Any practice not designed to meet a course training objective. Examples of maltraining include, but are not limited to, using abusive, excessive physical exercise or unnecessarily rearranging the property of an Airman to correct infractions. Any practice for the purpose of inducing an Airman to self eliminate [quit] is considered maltraining. (Air Education and Training Command Instruction 36-2216, 2010, p. 30)

Section II of the survey is designed to assess maltreatment and maltraining. The survey includes 17 items developed specifically to match BMT policy. The items are based on a review of AETC policy and MTI training materials, an AETC subject matter expert (SME) review of initially drafted items, feedback from MTIs and trainees, and item performance on the test survey (see Appendix A for more detail). The section includes five items focused on maltraining, two items focused on privacy violations, two items focused on denial of services or rights, two items focused on hostile comments, five items focused on physical threats or force, and one item focused on the encouragement of mistreating other trainees.

Section III: Unprofessional Relationships Between MTIs and Trainees

One impetus for the current survey was the discovery of unprofessional relationships occurring between MTIs and trainees. Unprofessional relationships are defined in Air Force policy as the following:

> Relationships are unprofessional, whether pursued on or off-duty, when they detract from the authority of superiors or result in, or reasonably create the appearance of, favoritism, misuse of office or position, or the abandonment of organizational goals for personal interests. Unprofessional relationships can exist between officers, between enlisted members, between officers and enlisted

members, and between military personnel and civilian employees or contractor personnel. Fraternization is one form of unprofessional relationship and is a recognized offense under Article 134 of the Uniform Code of Military Justice (UCMJ). (Air Force Instruction 36-2909, 1999, p. 2)

Unprofessional relationships in BMT specifically are described as well:

> The integrity and leadership of the faculty and staff in basic military training and the initial technical training environment must not be permitted to be compromised by personal relationships with trainees. At a minimum, faculty and staff will not date or carry on a social relationship with a trainee, or seek or engage in sexual activity with, make sexual advances to, or accept sexual overtures from a trainee. In addition, faculty and staff will not use grade, position, threats, pressure or promises to attain or attempt to attain any personal benefit of any kind from a trainee, or share living quarters with, gamble with, lend money to, borrow money from or become indebted to, or solicit donations (other than for Air Force approved campaigns) from a trainee. The same limitations govern personal relationships between faculty and staff and a trainee's immediate family members. Trainees have an independent obligation not to engage in these activities with members of the faculty and staff. (Air Force Instruction 36-2909, 1999, p. 4)

Section III of the survey is designed to assess unprofessional relationships. Like the maltreatment and maltraining section, the survey includes 16 items focused on unprofessional relationships developed specifically for the BMT context based on a review of AETC policy and MTI training materials, an AETC SME review of initially drafted items, feedback from MTIs and trainees, and item performance on the test survey (see Appendix A for more detail). The section includes five items focused on attempts to establish a personal or intimate relationship (e.g., talk about sex life), two items focused on inappropriate exchanges of money, three items focused on inappropriate social contact (e.g., inviting a trainee to a social gathering), and six items focused on other relationship policy violations (e.g., drinking alcohol with a trainee).

Section IV: Sexual Harassment of Trainees by Anyone at BMT

Civilian law and the UCMJ prohibit sexual harassment. Civilian legal precedent has established relatively clear workplace guidelines prohibiting both coerced sexual exchanges (quid pro quo) and offensive sex-related behavior that create a hostile environment. Quid pro quo behaviors extort sexual cooperation in exchange for job-related considerations (e.g., promising a promotion in exchange for a sexual encounter). Hostile environment behaviors are typically split between two conceptually distinct dimensions. The first, *gender discrimination*, includes a broad range of behaviors intended to convey insulting, hostile, or degrading attitudes toward women (e.g., describing men as "girls" or "sissies" as an insult). The second, *unwanted sexual attention*, includes sexual language and behaviors that are offensive, unwanted, and nonreciprocated (e.g., sharing explicit sexual stories with a coworker who has indicated discomfort). The UCMJ also prohibits sexual harassment, which includes "influencing, offering to influence, or threatening the career, pay, or job of another person in exchange for sexual favors, and deliberate or repeated offensive comments or gestures of a sexual nature" (United States Code, 2012, p. IV-26).

Section IV of the survey is designed to assess sexual harassment. The section includes 16 items, adapted to the BMT context, from an established measure of sexual harassment previously adapted to the military context. It is generally known as the Sexual Experiences Questionnaire-DoD (SEQ-DoD; see Stark et al., 2002). The section requests that respondents include behaviors that come from another trainee, an MTI, or others at BMT. Therefore, these items ask trainees whether "anyone" at BMT has behaved that way (the survey then includes follow-up questions to assess the characteristics of the offender). The items to assess sexual harassment include three focused on sexist hostility (behaviors that demean a person because of his or her gender), four focused on sexual hostility (behaviors in which sexual content is used purposely to offend a coworker), four focused on sexual coercion, and three focused on unwanted sexual attention. Although not included as part of the original SEQ-DoD, the survey also includes two additional items focused on challenges to masculinity/femininity. These are items that, on average, men are more likely to define as sexual harassment ("called you gay as an insult" and "insulted you by saying you were not acting like a real man or real woman") (Stockdale, Visio, and Batra, 1999). Because the SEQ-DoD was originally designed to assess sexual harassment of women, domains that men, on average, find most offensive are underrepresented.

It is important to note that sexual harassment scale scores from this measure must be interpreted as *experiences consistent with sexual harassment* rather than *sexual harassment* per se. Sexual harassment is a complex legal construct, and many scale items fail to include all indicators necessary to meet the legal standard (e.g., victim must be offended and offensive behavior must meet a "reasonable person" standard, or failing that, the perpetrator must be aware that the victim is offended and *continue* his or her behavior after learning that the behavior is offensive).

Follow-Up Questions

Any trainee who answers affirmatively to at least one item indicating an experience consistent with sexual harassment receives a series of follow-up questions that request details about the most serious event or the event that had the greatest effect on him or her. Although trainees may have experienced more than one event, asking details on each of the events listed would substantially lengthen the survey and could potentially identify victims to other trainees in the room because they would have a much longer completion time. Moreover, a single behavior often does not rise to the level of perceived sexual harassment; it is the cumulative effect of repeated behaviors that become problematic. To improve recall, some survey instruments focus respondents on the most recent event only for follow-up questions. For this survey, which prompts recall of eight weeks of experiences only, we did not believe that memory challenges would pose a significant threat. Additionally, we believed that AETC leadership would have a greater need for documentation of the most-serious events. Therefore, we chose to focus on the most-serious event in the follow-up questions.

These items were modeled after a similar set of items included in the Defense Manpower Data Center (DMDC) Workplace and Gender Relations Survey (Rock et al., 2011). Given the unique features of the BMT environment, we modified the original DMDC items and responses for this survey. Follow-up items assess the number of perpetrators ("one person" or "more than one person") and the gender and the status of the perpetrator(s) (trainee, MTI, other military personnel, or nonmilitary personnel).

Section V: Trainee Unwanted Sexual Experiences with Anyone at BMT

Sexual assault may be defined narrowly, by limiting the definition to completed vaginal rapes, or broadly, by including all forms of unwanted or coercive sexual contact. DoD takes the broader perspective and defines sexual assault as

> intentional sexual contact, characterized by use of force, threats, intimidation, abuse of authority, or when the victim does not or cannot consent. It includes rape, forcible sodomy (oral or anal sex), and other unwanted sexual contact that is aggravated, abusive, or wrongful (to include unwanted and inappropriate sexual contact), or attempts to commit these acts. (U.S. Department of Defense Directive 6495.01)

To reflect this broader definition and to avoid confusion, we use the term *unwanted sexual experience* to describe the content of this section.

Section V of the survey assesses unwanted sexual experiences using eight items developed specifically for this survey. Although there are several different established measures of sexual assault in the scientific literature, after a careful review, we decided that none was fully appropriate for the BMT context (see Appendix A for a review of other measures). We drew on the language in previously established scales to develop a new measure that included categories of sexual assault described in behaviorally specific terms and items consistent with the UCMJ definition of sexual assault.[5] The section includes one item to assess exposure of private areas of the body (added at the request of AETC); one item to assess unwanted sexual contact or frotteurism; three items to assess attempted oral, vaginal, and anal rape; and three items to assess completed oral, vaginal, and anal rape. With the exception of two items assessing vaginal

---

[5] Measurement strategies that define sexual assault narrowly, surveys that rely on crime reports, and surveys that use the word *rape* tend to produce small prevalence estimates, while those that ask behaviorally specific questions, which define events that meet the legal definition of sexual assault, tend to produce the largest prevalence estimates (Fisher, 2009; Tjaden and Thoennes, 1998). Given limited evidence that overreporting of sexual assault is common, combined with victims' disinclination to reveal sexual trauma, most investigators have urged reliance on measurement strategies that encourage accurate and full reporting. That is, estimates drawn from participants who were assured confidentiality and who responded to behaviorally specific items are considered by many to be more-accurate estimates of population incidence and prevalence.

assaults (which are not administered to male trainees), all items are equally applicable to both male and female victims. Male victims are less likely than female victims to report the assault via official channels (DoD, 2013); therefore, ensuring that the survey assesses victimization among men and women is vital to accurate tracking of all sexual assaults. Finally, given concern that many victims of sexual assault do not label the event *rape*, we chose to avoid this language, which may lead to underreporting. Instead, we chose the phrase *unwanted sexual experience* for survey items, which would capture incidents in which the trainee indicated verbal or physical nonconsent and also assaults in which nonconsent could not be communicated.

Unlike the other abuse and misconduct domains, we did not ask participants to indicate the frequency with which an unwanted sexual experience occurred, but only whether it occurred (yes or no). For trainees who respond in the affirmative, there is a follow-up question asking how many times they had an unwanted sexual experience. Finally, like the measurement of sexual harassment behaviors, these items ask trainees to respond whether "anyone" at BMT did this. The survey again includes follow-up questions to assess the characteristics of the perpetrator.

Follow-Up Questions

Any trainee who answers affirmatively to at least one scale item receives a series of follow-up questions that request details about the most serious event or the event that had the greatest effect on him or her. We chose to focus on the most serious event for the same reasons described in the previous section on sexual harassment follow-up questions. We also expect that due to the short duration of the training period, the proportion of trainees who experience multiple unwanted sexual experiences will be small, and, therefore, for most victims, follow-up questions will be easily mapped onto the single assault that they experienced.

These items were modeled after a similar set of items included in the DMDC Workplace and Gender Relations Survey (Rock et al., 2011). Given the unique features of the basic training environment, we modified the original DMDC items and responses for this survey. One follow-up item assesses the tactics used by the perpetrator. Some response options would indicate an incident that was consistent with coercion, but not a UCMJ-defined sexual assault ("showed displeasure, criticized my sexuality or attractiveness, or became angry"), while other response options are consistent with an assault ("threatened me with a weapon"). We included a range of possible tactics to allow leadership to track a variety of unwanted sexual encounters in the BMT environment. It will be important to report survey responses in the context of the tactic used to ensure that events are appropriately categorized. Follow-up items also assess the location of the incident (e.g., dorms, classroom, outdoors), the number of perpetrators ("one person" or "more than one person"), the gender and the status of the perpetrator(s) (trainee, MTI, other military personnel, or nonmilitary personnel), and support services received following the event (e.g., help from the sexual assault response coordinator [SARC], support from a chaplain).

Trainee Reporting or Telling Others About Abuse and Misconduct

Individuals who experience negative events such as bullying, sexual harassment, or unwanted sexual experiences may face considerable barriers to reporting those events. Victims may be uncomfortable making a report, do not want anyone to know about the incident, or fear that their confidentiality will not be protected (Rock et al., 2011). In a recent survey of military personnel, among sexual assault victims who chose to report the assault, 62 percent experienced professional retaliation (e.g., denied promotion), social retaliation (e.g., ignored by coworkers), or administrative actions (e.g., placed on a medical hold; Rock et al., 2011). For these reasons, some researchers and victim advocates have argued that a victim who chooses not to report the perpetrator has made a rational choice in which he or she believes that the negative consequences associated with reporting outweigh the potential personal benefit (Herbert and Dunkel-Schetter, 1992; Ullmnan, Foynes, and Tang, 2011). At the same time, a system as a whole (e.g., the AETC training environment) benefits from victims who report their experiences. Only after a victim (or a confidante or witness) reports an event can leadership intervene to remove the perpetrator from the environment and implement policies to reduce the risk of future assaults. Thus, it is important to understand the barriers that prevent some victims from reporting negative events, and, for those victims who do report these events, the nature of their experiences with the reporting system.

We developed a reporting section of the survey to assess the decisions and experiences of any trainee who either disclosed or chose not to disclose that he or she had experienced bullying, maltreatment or maltraining, unprofessional relationships, sexual harassment, or an unwanted sexual experience. To assess bystander reporting, trainees who indicated that they were aware of another trainee who had experienced one of these events were also surveyed about their disclosure decisions and experiences. Items include an assessment of:

- whether or not the trainee told someone about the incident
- how the trainee disclosed the incident (e.g., told an MTI, told someone else in the chain of command, used the dorm hotline)
- for trainees who chose not to disclose the incident, the barriers to disclosure (e.g., "I didn't think I would be believed"; "I was afraid reporting might cause trouble for my flight")
- for trainees who chose to tell someone about the incident:
  - how seriously they felt their report was taken
  - what happened with the behavior after they disclosed it (e.g., continued or got worse)
  - what happened after they reported (e.g., "the person who did it tried to get even with me for reporting"; "the person I reported it to praised me for reporting")
  - if they had it to do again, whether they would still report the incident (yes or no).

This section was based on a similar set of items included in the DMDC Workplace and Gender Relations Survey (Rock et al., 2011). However, we tailored the items and the responses to fit the unique features of the basic training environment.

14

Trainee Perceptions of the Squadron Climate

Organizational climate focuses on how individuals experience and make sense of an organization. Specifically, organizational climate involves perceptions of policies, practices, and procedures that are rewarded, supported, and expected. The importance of the organizational context in understanding workplace phenomena is well established, and, in particular, research has indicated the importance of organizational climates in shaping behavior (see Ostroff, Kinicki, and Tamkins, 2003). For example, research on safety climates within organizations has found a relationship with the number of accidents that take place (Christian et al., 2009) and even whether individuals are likely to report an accident (Probst, Brubaker, and Barsotti, 2008). Research has also examined the existence of climates for sexual harassment and the extent to which individuals perceive the organization tolerates sexual harassment and implements related policies and procedures designed to prevent it (see, e.g., Culbertson and Rodgers, 1997; Hulin, Fitzgerald, and Drasgow, 1996; Williams, Fitzgerald, and Drasgow, 1999). A recent meta-analysis found that organizational climate for sexual harassment was the strongest predictor of whether sexual harassment took place in the organization (Willness, Stell, and Lee, 2007).

Drawing on other established measures of climate and paring down a longer list of tested items, the survey includes separate scales of four items developed specifically for this survey to assess perceptions of the squadron climate for each of the abuse and misconduct domains. Although we could have assessed perceptions of the broader BMT or Air Force climate, we chose to focus specifically on the squadron level, since each trainee flight operates within a squadron that oversees the flight/MTI leadership. The items focus on the extent to which squadron leaders enforce policies and encourage the reporting of incidents within each abuse and misconduct domain. Since the survey does not include information on the specific squadron for each trainee, it is not possible to identify problematic squadrons or assess whether these perceptions are shared across trainees within the same squadron. Instead, the survey focuses on measuring climate at the individual level, also known as *psychological climate* (Ostroff, Kinicki, and Tamkins, 2003). Example items include "squadron leaders make honest efforts to stop unprofessional relationships" and "squadron leaders encourage the reporting of unprofessional relationships" (five-point response scale ranging from "strongly disagree" to "strongly agree).

Section VI: Trainee Perceptions of BMT Feedback and Support Systems

Recommendations adopted by the AETC commander following the 2012 detection of increased sexual misconduct at AETC included increasing the visibility of supervisors, commanders, chaplains, and SARCs in the training environment (Rice, 2012). Increasing trainees' access to and familiarity with the individuals who are appropriate reporting or support channels was expected to facilitate reporting of sexual abuse and other misconduct. RAND researchers designed Section VI of the survey to serve two purposes. First, trainees are asked to indicate how easy it would be contact the following people if they wanted to talk to them about the problems mentioned in the survey: instructor supervisor, flight commander, squadron superintendent, first sergeant, director of operations, squadron commander, an MTI, chaplain, and SARC, as well as additional support people, such as a BMT doctor or nurse, mental health professional, or law enforcement officer. Second, to assess the success of efforts to increase the visibility of individuals to whom a report could be made, the survey includes an item to query whether the trainees would recognize those individuals listed above (excluding the BMT support personnel with whom they may legitimately rarely come into contact).

In addition, this section of the survey assesses trainees' perceptions of the available feedback systems at BMT. Twelve items meant to be analyzed individually (rather than as a scale) assess trainees' level of agreement or disagreement with a variety of statements, including, for example: "BMT makes it easy to use a dorm hotline," and "if I experienced abuse or mistreatment from a MTI, there is at least one person at BMT in the chain of command I feel I could turn to for help." These items are designed to assist AETC headquarters and BMT leadership in evaluating the success of efforts to restore trust and improve the likelihood that trainees will report an incident of misconduct.

Closing Question

The survey ends with a single question asking trainees how open and honest they felt that they could be when answering the survey questions. This question serves three purposes. First, it can serve as an indicator of trust in BMT leadership and the organization as a whole. Second, it provides an assessment of the effectiveness of the administrative procedures for the survey. If a large number of trainees were not open and honest on the survey, AETC should examine what improvements can be made to promote perceptions of greater confidentiality and trust. Third, the question serves as a way to screen out participants who were not answering in an accurate manner, which could potentially bias results.

What This Trainee Survey Excludes and Why

In developing the survey, RAND considered the inclusion of a number of different topics based on reviews of AETC and DoD reports and policies, meetings with leaders, feedback from MTIs and trainees, a test of the survey instrument, and a review of the scientific literature. The survey in this report represents RAND's efforts to address the key goals of the survey (help detect incidents of abuse and misconduct and provide data to help leaders understand what actions to take to reduce abuse and misconduct) while balancing the survey length to avoid trainee fatigue. We considered three key areas for inclusion in the survey but ultimately recommended excluding them: (1) an evaluation of BMT training related to abuse and misconduct, (2) more-detailed feedback from victims of sexual assault, and (3) questions asking trainees if they were perpetrators of abuse or engaged in misconduct.

As part of BMT, trainees receive educational training on the different types of abuse and misconduct assessed in this survey. Evaluating whether this training is effective, whether trainees understand what constitutes different types of abuse and misconduct (e.g., maltreatment, maltraining, and unprofessional relationships), and what to do about them is an important part of a comprehensive prevention effort. However, since the survey is already fairly lengthy and a training evaluation would not need to occur as frequently as the planned survey administration (i.e., a training evaluation could occur only every few years unless the training changes or there is a change in the number of incidents or lack of reporting), we recommended excluding these types of questions from the current trainee survey. Instead, we recommend that AETC pursue a separate training evaluation.

Similarly, we also considered including more-detailed questions on the experiences of individuals who reported an unwanted sexual experience, such as the quality of care they received, treatment by law enforcement and the judicial system, and squadron leaders' reactions. However, based on published estimates of sexual assault across the Air Force (Rock, 2013) and the size of the trainee population in each administration cycle, the number of victims will likely be zero or very small each administration cycle, or even each quarter. Even at the height of the sexual misconduct that preceded this study, the number of victims at any point in time was relatively small from the perspective of a viable survey sample size. We were concerned that the inclusion of the additional items could risk identifying the victims of unwanted sexual experiences when results were disseminated. We were also concerned that the additional time victims would need to complete the survey might be revealing to other trainees in the room. The anticipated small number of victims would also make it difficult to statistically analyze the data. Furthermore, some of these activities (postassault care, prosecution of perpetrators) could extend well beyond the BMT time frame. Based on these concerns, we recommended excluding these types of questions from the survey. Instead, we recommend that AETC explore other mechanisms for understanding victim experiences and care following reporting. Key domains to include in such an examination would be experiences with the reporting process, judicial system, victim care/advocacy system, and leadership reactions.

Finally, we also considered asking questions to assess the extent to which trainees may have committed acts of abuse or misconduct, but ultimately decided to exclude this assessment. The main purpose of developing the survey was to assess rates of abuse and misconduct toward trainees. Evidence suggests that victims are more likely to report incidents than perpetrators, and as a result, an incidence estimate computed from victims will be closer to the true incidence than a perpetrator-based estimate (Lewis and Fremouw, 2001). Thus, asking about perpetration of each of these behaviors would lengthen the survey considerably and likely result in a less accurate estimate. Therefore, given limited space, we decided it best to focus on victimization. To help develop a more comprehensive picture of sexual assault in the Air Force and potentially enhance prevention efforts, however, we do recommend that the Air Force undertake a thorough examination of the characteristics, motivations, and behaviors of perpetrators.

Analyzing and Interpreting Results from the Trainee Survey

Below, we provide an overview of recommendations for analyzing and interpreting results from the trainee survey. As one of the key deliverables for this study, RAND also developed a reporting template to help facilitate the recommended analyses and tracking trends over time. An illustrative snapshot of that template is included in Appendix D.

Preliminary Data Validation

To ensure high confidence in reported data, particularly reports of maltreatment or wrong doing, it is important to validate each trainee's data and to remove from the data set the responses of any trainee who did not complete the questionnaire in good faith. Embedded in each survey are items that will assist the data analyst in identifying these cases and isolating them from the final data set. At the end of the survey, each trainee responds to the item, "How open and honest did you feel you could be when answering these survey questions?" The responses of any trainee who indicates that he or she was "not at all open or honest" should be noted and then dropped from the final data set, as these admittedly dishonest responses may contaminate the full data set. Given the possibility that these individuals may have been less likely to be honest due to lack of trust, fear of repercussion, or the experience of abuse at BMT, however, we suggest examining their responses separately and noting this unknown quantity when providing results. Additionally, each of the five main sections includes an item to assess whether trainees are reading and responding to the questions carefully. For example, Section I includes the following: "Please select 'Daily' for this item to help us confirm that trainees are reading these items." A trainee who is responding quickly without reading items (e.g., checking "never" for every question) may not correctly respond to this item and thus can potentially be identified as someone for whom confidence in the accuracy of his or her responses should be low. Five items in the questionnaire check if trainees are reading and responding carefully (1.1e, 2.1k, 3.1n, 4.1k, 5.1e). If a respondent answers any of these items incorrectly, he or she should be excluded from the final data set.

Prevalence of Abuse and Misconduct

To examine the prevalence of each type of abuse and misconduct (bullying, maltreatment and maltraining, unprofessional relationships, sexual harassment, and unwanted sexual experiences), we recommend that the data analyst summarize the data in the following different ways. First, at a very high level, an overall summary variable can be created to describe the number and percentage of trainees who endorse *experiencing at least one incident* within the particular domain while at BMT. This dichotomous variable combines trainees who indicate experiencing only one behavior on a limited basis with those who may have experienced repeated or more-serious behaviors. These results, and all of the other prevalence results, should be reported by gender.

18

Second, to better account for the differences between these extremes, we also recommend that experiences be summarized as a frequency variable (reporting number and percentage of trainees) representing the greatest frequency with which trainees indicate experiencing any behavior. The frequency variable should group trainees into five different categories: those who report experiencing (1) no incidents, (2) incidents once or twice *at most*, (3) incidents a few times *at most*, (4) incidents on a weekly basis *at most*, or (5) daily incidents. Note, for unwanted sexual experiences, participants were only asked whether the incident happened to them (yes or no). Therefore, this frequency variable does not apply for analyzing this domain on the survey.

In order to have a better understanding of the specific types of abuse or misconduct taking place within each domain, we also recommend summarizing the number and percentage of trainees who indicate experiencing each behavior (i.e., creating a dichotomous variable for whether a trainee experienced an item). Because there are a large number of items in most of the abuse and misconduct domains, this can be done by first summarizing the number and percentage of trainees who indicated experiencing at least one item in identified subscales within each domain. The subscales represent conceptual groupings of similar behaviors. For example, all the unprofessional relationship items that relate to MTIs making personal or unofficial contact with trainees are grouped together. For the sexual harassment scale, which has been adapted from a validated scale in the scientific literature, there are already established subscales (sexist hostility, challenges to masculinity/femininity, sexual hostility, sexual coercion, and unwanted sexual attention). Specific subscales for each abuse and misconduct domain are included as part of the survey analysis in Appendix B. To drill down further, analysts may also wish to create the same dichotomous variable for each individual item.

For the bullying scale, it may also be valuable to create a summed scale score by summing trainee responses across all items. This type of analysis is consistent with recommendations for analyzing other bullying measures (see Hamburger et al., 2011), given that bullying is considered to involve intentional hurtful behaviors that are repeated over time. Higher scores represent greater victimization from bullying. Reports of the summed score can then be based on the number and percentage of trainees within six-point intervals—for example, none (i.e., trainees who experienced no incidents), limited (score of 7–12), some (score of 13–18), and so on.

Follow-Up Questions

Both the sexual harassment and unwanted sexual experiences sections of the survey contain follow-up questions for trainees who indicated experiencing an incident while at BMT. For these questions, we recommend summarizing the results by providing the number and percentage of trainees who endorse each response. These results should also be reported by gender to better understand the different experiences of male and female trainees.

However, given that only a small number of trainees in any training week may indicate having some of these experiences, particularly unwanted sexual experiences, it will be important to consider, prior to reporting the data, whether data from these follow-up questions risk identifying the trainee by inference. To avoid this potential breach of anonymity, we recommend that these data be reported only when they can be sufficiently aggregated to avoid identification by inference. This determination will need to be made by a qualified data analyst who carefully considers the nature of the follow-up question and how it will be reported in relation to other questions on the survey. Therefore, the data analyst may wish to report these variables in aggregate either quarterly or annually.

Trainee Reporting or Telling Others About Abuse and Misconduct

Data on trainees' official and unofficial reports of abuse and misconduct should be summarized for each domain separately (e.g., bullying, maltreatment and maltraining). This involves summarizing the number and percentage of trainees who reported the incident either officially or unofficially, including the different reporting channels used (e.g., to an MTI, via the dorm hotline, to a SARC). For those trainees who experienced a prohibited behavior, but who chose *not* to disclose the incident, we recommend summarizing the number and percentage who endorsed each reporting barrier. Similarly, for those trainees who did disclose, summarize the number and percentage indicating the various response options for each question about their reporting experience.

Trainee Perceptions of Squadron Climate

Each abuse and misconduct domain also includes four items to measure trainee perceptions of the squadron climate. These items are part of a unidimensional scale; responses to the items should be averaged together to create a single scale score for each respondent. The average score across trainees can then be reported to represent overall perceptions of the extent to which squadron leaders enforce policies and encourage the reporting of incidents, with higher scores indicating more-positive views. Of course, it will also be important to examine the variance in responses as well as potential differences in male and female perceptions.

Trainee Perceptions of BMT Feedback and Support Systems

Finally, Section VI of the survey includes items assessing BMT feedback and support systems. These items are meant to be analyzed separately by examining the number and percentage of trainees who endorse each response option.

Background Characteristics

We recommend that all of the above analyses be conducted separately for males and females. Men and women are likely to have different experiences at BMT, particularly related to abuse and misconduct. Understanding where these differences exist will help leadership better address issues and tailor prevention efforts.

Summary

This BMT trainee survey provides a framework for assessing the prevalence and related reporting behaviors in five core abuse and misconduct domains important to BMT. The survey also includes items to assess individual perceptions of the squadron climate in terms of the extent to which squadron leaders are perceived to enforce laws and policies related to each of the abuse and misconduct domains. A final section then assesses perceptions of the feedback and support systems available at BMT. Table 2.1 provides an overview of these domains and final survey content.

Table 2.1. Overview of BMT Trainee Final Survey Content

Survey Content	Number of Items
Background demographics	
Gender	1 item
Section I: Trainees bullying other trainees	
No subscales	6-item scale
Section II: Maltreatment and maltraining by MTIs	
Maltraining	5-item subscale
Privacy violations	2-item subscale
Denial of services or rights	2-item subscale
Hostile comments	2-item subscale
Encouragement of mistreating other trainees	1-item subscale
Physical threats or force	5-item subscale
Section III: Unprofessional relationships with MTIs	
Attempts to establish a relationship	5-item subscale
Inappropriate exchanges of money	2-item subscale
Inappropriate social contact	3-item subscale
Relationship policy violations	6-item subscale
Section IV: Sexual harassment from anyone at BMT	
Sexist hostility	3-item subscale
Sexual hostility	4-item subscale
Sexual coercion	4-item subscale
Unwanted sexual attention	3-item subscale
Challenges to masculinity/femininity	2-item subscale
Unique follow-up questions	3 separate items
Section V: Unwanted sexual experiences committed by anyone at BMT	
Exposure of private areas of the body	1-item subscale
Sexual contact	1-item subscale
Attempted rape	3-item subscale
Completed rape	3-item subscale
Unique follow-up questions	7 separate items
Trainee reporting or telling others about abuse and misconduct (repeated in each abuse and misconduct section)	
Aware of other trainees experiencing abuse and misconduct	1 item
Told other trainees about behavior personally experienced	1 item
Reported abuse and misconduct to any Air Force authority	1 item
Reasons for not reporting	21 separate items
Experiences reporting	4 separate items
Trainee perceptions of squadron climate (repeated in each abuse and misconduct section)	
Bullying	4-item scale
Maltreatment and maltraining	4-item scale
Unprofessional relationships	4-item scale
Sexual harassment	4-item scale
Sexual assault	4-item scale
Section VI: BMT feedback and support systems	
Ease of contacting BMT personnel	12 separate items
Recognition of BMT personnel	8 separate items
Perceptions of available feedback systems	12 separate items
Closing question	
Open and honest	1 item

3. The MTI Survey

The MTI survey is a separate but complementary survey to the trainee survey. The MTI survey collects additional data on abuse and misconduct toward trainees at BMT. Like the trainee survey, the MTI survey is designed to help AETC headquarters and BMT leaders detect incidents of abuse and sexual misconduct and to provide data to help them develop appropriate actions to reduce both. Additionally, the MTI survey includes a section assessing overall MTI quality of life. Prior to the current project, AETC already had an MTI Quality of Life Survey (QOLS), which was administered to MTIs once a year. In order to minimize the risk of survey burnout among MTIs, AETC requested that RAND review the QOLS and determine whether the constructs it measured could be integrated into RAND's survey. We were able to integrate QOL constructs, and in many ways, the QOLS survey was already aligned with the purposes of the new abuse and misconduct survey. An MTI's job performance is defined not only by the ability to prepare trainees to become airmen but also by the extent to which maltraining, maltreatment, unprofessional relationships, and harassment of trainees are avoided. Moreover, MTIs are responsible for ensuring that their fellow MTIs also comply with these policies and report any known or suspected violations. Whether MTIs meet these expectations will be influenced to some extent by the attitudes they have toward the Air Force, BMT, trainees, and their jobs.

This chapter is dedicated to describing the content of the final survey RAND developed, which can be reviewed in full in Appendix C. We also describe the recommended analytic and reporting procedures for the survey in this chapter. For the interested reader, we present greater detail on the development of the survey in Appendix A.

MTI Survey Overview

The MTI survey is split into two main parts. The first part assesses MTI quality of life, including MTI job attitudes, perceptions of the work environment, job stressors, and MTI professional development. The second part of the survey is then designed to provide an assessment of the extent to which MTIs are *aware* of trainees experiencing the same abuse and misconduct behaviors addressed by the trainee survey. The survey also includes a section on MTI perceptions of squadron climate related to each abuse and misconduct domain, perceptions of MTI reporting norms, and the extent to which related policies are clearly defined. Based on a test of the survey, the estimated average time to complete the survey is 25 minutes. However, some MTIs may take considerably longer, particularly those who choose to write lengthy comments at the end of the survey.

MTI Survey Section Descriptions

Background: Demographics

The survey includes only two background questions to help ensure anonymity. First, since the length of time someone has been an MTI may influence his or her perceptions of the environment and job attitudes, the survey asks participants how long they have been an MTI in total (response options: six months or less; greater than six months, but less than two years; or two years or more). The second background question asks about the primary duty of the MTI, since differences in responsibilities and interactions with trainees may similarly influence attitudes and opportunities for abuse or misconduct (response options: line, supervisor, or other).[6]

There are certainly many other background questions that could be asked as part of the survey and show interesting differences (e.g., race/ethnicity, squadron, gender). However, the more background questions that are included, the greater the possibility of being able to identify an individual MTI based on unique combinations of background characteristics. For example, although we may find differences in perceptions of the environment between female and male MTIs, there are only a limited number of female MTIs at BMT (56 out of 490 as of June 2012; although BMT has been striving toward a goal of 25-percent female MTIs). Therefore, in order to encourage participation and protect respondents' identities, we chose to not include questions on gender or additional background characteristics in the survey.

Sections I–V: MTI Quality of Life

The first part of the MTI survey is designed to assess MTI quality of life, including MTI job attitudes, perceptions of the work environment, job stressors, and MTI professional development. It reflects an integration of the constructs measured on the original AETC MTI QOLS, as well as items designed to measure other topics raised in AETC reports and during meetings with MTIs and AETC headquarters and BMT leaders.

Section I: Job Attitudes

Job attitudes have been shown to be useful indicators of both employees' well-being and their job performance (Cooper-Hakim and Viswesvaran, 2005; Meyer et al., 2002; Meyer and Maltin, 2010; Riketta, 2002). Items measuring commitment were already included on the QOLS, and we

---

[6] Although the majority of MTIs serve as line instructors who directly oversee and are responsible for trainee flights, there are also MTIs who serve as instructor supervisors, responsible for overseeing the line MTIs and trainee flights within a squadron, and other MTIs who may be in a support role, work as an instructor (e.g., War Skills and Military Studies), or oversee the field training exercise that all trainees must complete at the end of BMT (Basic Expeditionary Airman Skills Training exercise [BEAST]).

concluded that organizational attitudes would be consistent with AETC's broader goals of creating a positive and safe work environment for MTIs and trainees.

Organizational Commitment

Commitment is seen as an important construct to monitor in organizations, as it provides a force that guides employee behavior and "binds the person to a course of action" (Meyer and Herscovitch, 2001, p. 301). Although there are different conceptualizations of organizational commitment (e.g., Jaros et al., 1993; Mayer and Schoorman, 1992; O'Reilly and Chatman, 1986), most overlap to some extent with the three-component model of organizational commitment developed by Allen and Meyer (1990). This model distinguishes among three well-known facets of commitment: affective, normative, and continuance commitment. *Affective commitment* refers to an emotional attachment, sense of loyalty, and identification with the organization and its values. *Normative commitment* is the extent to which employees remain with an organization and pursue a course of action out of a sense of duty, responsibility, and obligation to the organization. Finally, *continuance commitment* refers to the extent that employees continue to stay with an organization because the relative costs of leaving are too large when compared with the amount of investments they have made within the organization.

Extensive research conducted on these facets shows that affective commitment demonstrates the strongest and most-consistent links with desired organizational outcomes, including attendance, performance, and organizational citizenship behavior (Meyer, Stanley, et al., 2002). This research has also shown that affective commitment is moderately related to stress, role conflict, and work-family conflict. In a sample of Army captains, affective commitment was also shown to predict peer ratings of leadership (Karrasch, 2003). Normative and continuance commitment generally demonstrate weaker relationships with these criteria; therefore, we considered only items measuring affective commitment for inclusion in the MTI survey.

To assess affective commitment, the survey includes a five-item scale, with the items adapted from a well-established scale of organizational commitment (Allen and Meyer, 1990; Meyer, Allen, and Smith, 1993). Example items include "BMT has a great deal of personal meaning for me" and "I would recommend becoming an MTI to others" (five-point response scale ranging from "strongly disagree" to "strongly agree").

Job Satisfaction

Job satisfaction can be thought of as "a pleasurable or positive emotional state resulting from the appraisal of one's job or job experiences" (Locke, 1976, p. 1300). Although job satisfaction is not as clearly linked to job performance (Bowling, 2007), it can serve as an important indicator of an employee's health and well-being (Bowling, Eschleman, and Wang, 2010; Faragher, Cass, and Cooper, 2005). Job satisfaction is also an important factor for retaining qualified employees (Griffeth, Hom, and Gaertner, 2000; Tett and Meyer, 1993). Therefore, efforts to track and maintain high levels of job satisfaction should be key objectives to ensure that the best and most-qualified airmen not only serve as MTIs but continue to stay in the Air Force.

To assess job satisfaction, the survey includes a one-item global measure of job satisfaction, as well as several facet measures of job satisfaction, which can be more diagnostic of potential problems within an organization. Global measures of job satisfaction may indicate that employees are generally unhappy but do not provide organizational leaders with the diagnostic information necessary to target specific areas for improvement (e.g., coworker conflict). Therefore, we included both approaches to measuring job satisfaction as part of the survey. To assess global job satisfaction, the survey includes the following well-established one-item measure: "Considering everything, how would you rate your overall satisfaction with your job as an MTI?" (Highhouse and Becker, 1993). The survey assesses facet measures of job satisfaction most relevant to MTIs with the scales described in Section II.

Section II: Work Environment

Organizational Support

Having sufficient organizational support in terms of resources and training is critical to effective performance, with meta-analyses finding that insufficient resources to meet work demands can often lead to work overload and affect satisfaction, stress, and performance (Gilboa, Fried, and Cooper, 2008; Spector and Jex, 1998). Therefore, the extent to which workers feel that they have the organizational support or resources needed to do their jobs effectively is important to monitor. To assess the potential resource constraints that MTIs might experience, the survey includes four items meant to be analyzed separately, which were developed specifically for this survey. Example items include "I have the resources or equipment I need to carry out my job duties effectively" and "there are not enough MTIs to carry out all MTI duties effectively" (five-point response scale ranging from "strongly disagree" to "strongly agree").

MTI Interpersonal Treatment

Our reviews of AETC reports and meetings with leaders and MTIs indicated that the interpersonal treatment among MTIs may be an issue as it relates to fostering a positive and teamwork-focused environment at BMT. Further, recent research on interactions among coworkers, particularly the extent to which individuals feel that they are treated fairly and with respect has been found to affect key work attitudes and team processes, including workplace aggression (e.g., Cropanzano, Li, and Benson, 2011; Hershcovis et al., 2007; Li and Cropanzano, 2009). In the BMT environment, interactions among MTIs are further complicated by a distinct power differential between seasoned MTIs and rookie MTIs.[7] Therefore, to measure the interpersonal treatment among MTIs, the survey includes two four-item scales adapted from an established measure of interpersonal treatment (Donovan, Drasgow, and Munson, 1998). The first scale assesses how well seasoned MTIs treat rookie MTIs at BMT, and the second scale then assesses how well MTIs treat each other more generally. Example items include "MTIs help each other out" and "MTIs treat each other with respect" (five-point response scale ranging from "strongly disagree" to "strongly agree").

MTI Perceptions of Trainees

AETC reports and MTI reviews of an initial survey draft indicated some concern that trainees may have too much power over how MTIs perform their duties. Specifically, concerns were raised that trainees have less respect for MTIs than in the past and that leaders trust trainees more than MTIs. To monitor MTI concerns about trainees, the survey includes six items meant to be analyzed individually (rather than as a scale), which were developed specifically for this survey. Example items include "leaders take the word of trainees over the word of MTIs" and "trainees respect MTI authority" (five-point response scale ranging from "strongly disagree" to "strongly agree").

Assignment and Promotion Opportunities

This section asks MTIs to reflect on MTI assignment and promotion opportunities. This is consistent with other facet measures of satisfaction (e.g., Kinicki et al., 2002) and may affect the extent to which BMT is able to attract the best and brightest to become MTIs. The section includes two items designed to be analyzed separately. One item assesses perceptions of assignment opportunities ("Those who do well as an MTI are given fair consideration for a good follow-on assignment") and one item assesses perceptions of promotion opportunities ("MTIs are promoted to the next higher rank more slowly than NCOs serving in other assignments"; five-point response scale ranging from "strongly disagree" to "strongly agree").

Section III: Leadership

Leader Treatment of MTIs

Research has shown that a leader's treatment of subordinates, such as whether individuals are treated fairly and with dignity and respect, can have a significant impact on their work attitudes and performance (Cohen-Charash and Spector, 2001; Colquitt et al., 2001), including workplace aggression (Hershcovis et al., 2007). To measure leader treatment of MTIs, the survey includes a ten-item scale adapted from the same established measure used to measure MTI interpersonal

---

[7] *Rookie* refers to newer, less-experienced MTIs, and there is no formal specific set cutoff point between rookie and seasoned MTI. This distinction is based on general MTI perceptions. The terms *rookie* and *seasoned* were recommended by BMT staff to capture the distinction between those with and without seniority/experience.

treatment (Donovan, Drasgow, and Munson, 1998). The scale includes ten items asking MTIs to rate their agreement regarding whether their immediate supervisor engages in different behaviors. Example items include "plays favorites" and "trusts MTIs" (five-point response scale ranging from "strongly disagree" to "strongly agree").

Leader Ethical Conduct

The Air Force expects all airmen to act ethically and uphold the three Air Force core values of "integrity first," "service before self," and "excellence in all we do." Airmen in leadership roles have an even greater obligation of modeling and providing guidance on how to uphold these values and behave ethically as airmen. A study with Army soldiers found that ethical leadership was positively related to the existence of ethical culture and soldiers' ethical behaviors, with leaders at higher organizational levels having a cascading effect on the culture that developed in lower units (Schaubroeck et al., 2012). To assess MTI perceptions of the ethical behavior of leaders in BMT, the survey includes a five-item scale adapted from an established measure of ethical leadership (Brown, Trevino, and Harrison, 2005). The items ask MTIs to rate their agreement regarding whether their immediate supervisor engages in different behaviors. Example items include "conducts his/her personal life in an ethical manner" and "sets an example of how to do things the right way in terms of ethics" (five-point response scale ranging from "strongly disagree" to "strongly agree").

Section IV: Work and Family Stressors

The survey also incorporates measures to address factors that may influence the stress experienced by MTIs. Stress is an important construct to monitor; not only does it serve as a strong indicator of an MTI's quality of life and well-being but high stress levels can deplete important self-regulatory resources needed to manage emotions and behavior in the BMT environment. For example, research has shown that organizational stressors can contribute to experiencing negative emotions, which then lead to increased levels of aggression and counterproductive work behavior (e.g., interpersonal conflict, theft, withdrawal) (Fox, Spector, and Miles, 2001). Other research from a meta-analysis has also shown that different forms of stress (e.g., conflict, work overload) are negatively related to organizational citizenship behaviors, such as taking on extra assignments (Eatough et al., 2011).

The survey has two separate sections designed to measure stress. The first section addresses the extent to which an MTI's work and family life impact each other. The second section attempts to identify other specific work-related stressors that affect MTIs. The survey does not attempt to capture every possible source of stress for MTIs. Because the MTI survey is already lengthy, we advise against extending the list unless a truly unique stressor not captured by the survey emerges as a significant concern.

Work-Family Conflict

Work-family conflict is often defined as "a form of inter-role conflict in which the role of pressures from the work and family domains are mutually incompatible in some respect" (Greenhaus and Beutell, 1985, p. 77). This was raised as a particular concern in our reviews of AETC reports, as well as MTI reviews of prior survey drafts. To assess the extent to which MTIs experience work-family conflict, the survey includes a six-item scale adapted from an established work-family conflict scale (Carlson, Kacmar, and Williams, 2000). The scale is a two-dimensional scale, with three items measuring strain-based family interference with work (e.g., "Due to stress at home, I am often preoccupied with family matters at work") and three items measuring strain-based work interference with family (e.g., "When I get home from work, I am often too worn out to participate in family activities or responsibilities"). The full scale contains four other dimensions of work-family conflict focused on time and behavior interference. However, because many of the items in those subscales were not as relevant for the BMT environment or were already assessed elsewhere on the survey, we chose to focus only on the strain-based interference subscales in the survey. The items use a five-point response scale (ranging from "strongly disagree" to "strongly agree"), which also includes the option of selecting "not applicable" for those individuals who feel that a particular item is not currently relevant (e.g., someone who is single and childless may not have family obligations at home).

Specific Stressors

This section asks MTIs to identify specific sources of job stress in their lives. The purpose of this section is to help leaders diagnose areas of concern and develop plans to improve MTIs' quality of life and performance, and to foster a safe and productive environment for MTIs and trainees. Based on a review of AETC reports, previous Air Force surveys, meetings with AETC leaders and MTIs, and written comments on a test of the survey, the survey includes a total of 28 potential work-related stressors (e.g., inconsistent policy guidance, lack of time off, long work hours) and asks MTIs to indicate the extent to which each has caused stress over the past six months (five-point response scale ranging from "no stress" to "a great deal of stress"). These items are not intended to form a single scale score representing the average level of stress across items; instead, they are designed to be analyzed separately to allow leadership to identify which areas are causing the most stress for MTIs. Although constructed to represent areas of particular concern to BMT, many of these stressors have also been highlighted in research in both military and civilian contexts. For example, several stressors reflect different forms of role stress, including role ambiguity and role conflict (e.g., conflicting job expectations). In contrast, several other stressors are unique to BMT (e.g., trainee comment forms).

Two follow-up questions also ask MTIs about their work and sleep habits. Although the section on specific stressors includes items that assess MTI stress related to being overworked, we also sought to obtain more-quantitative measures of the amount of time MTIs work and are able to sleep. The survey includes a single question on the average number of hours worked in a day, as well as a single item on the number of hours of sleep the MTI is able to obtain in a 24-hour period and a single item on his or her overall sleep quality.

Section V: MTI Professional Development

The professional development section of the survey includes two questions taken directly from the MTI QOLS. AETC considered answers to these questions to be important to continue to track in order to have feedback relevant to its MTI development programs. The first question asks MTIs about the extent to which they believe their instructional skills have improved over the past six months (five-point response scale ranging from "to a great extent" to "to no extent"). The second question asks MTIs if they have taken a deliberate development course since becoming an MTI and the extent to which they felt that the course was beneficial (four response options: "Yes, and I benefited from taking it"; "Yes, but it wasn't very helpful"; "No, but I would like to"; "No, and I'm not interested").

Sections VI–XI: Abuse and Misconduct Toward Trainees

Awareness of Bullying, Maltreatment and Maltraining, Unprofessional Relationships, Sexual Harassment, and Unwanted Sexual Experiences

To complement the trainee survey, the MTI survey includes scales to assess MTI *awareness* of trainees bullying other trainees, MTI maltreatment and maltraining of trainees, unprofessional relationships between MTIs and trainees, sexual harassment of trainees, and trainees' unwanted sexual experiences in the BMT environment. The content of these scales mirrors the content of the trainee survey described in the previous chapter. RAND researchers considered assessing MTIs' self-reported engagement in misconduct, but, on the basis of feedback from MTIs, AETC headquarters, and BMT leadership, we determined that no matter the safeguards, it was highly unlikely that MTIs would feel sufficiently protected to reveal incidents of personal misconduct. To determine the prevalence of MTI abuse and misconduct toward trainees, we determined that the best approach would be to use trainees' reports of victimization. Thus, instead of assessing MTIs' personal misconduct, these sections ask MTIs to report whether they are personally aware of specific incidents of misconduct occurring at BMT during the past six months. These survey sections should be considered an assessment of the overall environment (MTIs' perceptions of abuse and misconduct occurring toward trainees). In addition, the extent of match or mismatch between MTI awareness of misconduct and trainee reports of misconduct can be considered an indirect measure of the success of the reporting systems.

With the exception of minor wording changes to shift the focus of items from trainee self-report to MTI reports of their awareness of these events happening to trainees, the scales assessing bullying, maltreatment and maltraining, unprofessional relationships, and sexual harassment are the same for the trainee and MTI surveys. The scale assessing trainee unwanted sexual experiences was modified for the MTI survey. Training materials for MTIs indicate that MTIs should not press trainees who report a sexual assault for a detailed description of the assault, but rather should focus on obtaining appropriate support and services for the trainee. As such, RAND researchers did not believe that MTIs, even those who had handled a report of an unwanted sexual experience, would necessarily be aware of accurate and precise details of the assault (e.g., whether the assault was oral, vaginal, or anal). Thus, we reduced the MTI unwanted sexual experiences scale to four items that assessed whether MTIs were personally aware of the following types of unwanted trainee sexual experiences occurring: exposure of private areas of the body; unwanted sexual contact or frotteurism; a single-item to assess oral, vaginal, or anal sex assault; and a single-item assessing incidents of attempted oral, vaginal, or anal assault (response options: "No, I'm not personally aware of this happening"; "Yes, I am personally aware of this happening").

MTI Perceptions of Squadron Climate

In each of the abuse and misconduct sections, the MTI survey also includes scales to measure MTI perceptions of the squadron climate or the extent to which MTIs perceive that squadron leaders enforce policies and encourage the reporting of incidents related to the particular abuse and misconduct domain. The content of these scales mirrors the content of the trainee survey, which is described in detail in Chapter Two.

MTI Reporting Norms

To assess potential barriers to MTI willingness to report abuse and misconduct, the survey includes four items meant to be analyzed separately for each of the abuse and misconduct domains. These items ask participants to think about MTI behavior in general at BMT and to indicate the extent to which they agree or disagree with a series of statements regarding MTI reporting. Example items include "MTIs would report another MTI for maltreatment or maltraining" and "an MTI who reported another MTI for maltreatment or maltraining would experience retaliation from other MTIs" (five-point response scale ranging from "strongly disagree" to "strongly agree"). The only abuse and misconduct domain that does not include these items is bullying, since MTIs are expected to personally address bullying among trainees as part of their duties. Unlike the other abuse and misconduct domains, bullying would also not require an MTI to make a report against a fellow MTI, so the hurdles of reporting a peer or superior are not present with trainee misconduct.

RAND researchers considered assessing MTIs' self-reported actions of reporting or not reporting incidents of which they were aware and potential barriers for why they chose not to report an incident. However, on the basis of feedback from MTIs, AETC headquarters, and BMT leadership, we determined that it was unlikely that MTIs would feel sufficiently protected to reveal that they failed to report an incident when it is considered part of their job duty. Therefore, we determined that the best approach would be to ask MTIs to report whether they perceived *MTIs in general* being willing to report an incident. This provides an assessment of perceptions

31

of the general reporting environment or norms for reporting behaviors instead, which is likely to influence individual MTI reporting actions.

Clarity of Abuse and Misconduct Policies

Section XI of the survey includes five separate items designed to assess how clear are the policies related to each of the survey's abuse and misconduct domains (five-point response scale ranging from "strongly disagree" to "strongly agree"). Based on meetings with AETC headquarters, BMT leaders, and MTIs, a concern was raised that policies, such as the types of discipline tools that MTIs are allowed to use, can be unclear or inconsistently enforced. A first step in ensuring that MTIs behave in accordance with leadership expectations is that they have the necessary knowledge to understand what is expected of them. Therefore, these items are intended to assess the extent to which relevant policies and laws are clearly understandable to MTIs. These items are meant to be analyzed separately.

Closing Questions

The final section of the survey ends with two questions. The first question asks how open and honest MTIs felt that they could be when answering the survey questions. This question serves three purposes. First, it can serve as an indicator of trust in BMT leadership and the organization as a whole. Second, it provides an assessment of the effectiveness of the administrative procedures for the survey. If a large number of MTIs indicate they felt that they could not be open and honest on the survey, AETC will need to examine what improvements can be made to promote perceptions of greater anonymity and trust. Third, the question serves as a way to screen out participants who were not answering in an accurate manner, which could potentially bias results.

The second question is an open-ended question that provides an opportunity for MTIs to communicate any additional issues or feedback they would like to share about their working conditions, leadership, or quality of life. The survey cannot assess every topic of MTI quality of life, and new issues or concerns may arise over the years. Therefore, this open-ended question provides an avenue for MTIs to voice any additional concerns or issues not assessed in the survey. To the extent that certain issues are voiced continuously, AETC may wish to consider adding those topics to future versions of the MTI survey.

What This MTI Survey Excludes and Why

In developing the MTI survey, RAND considered the inclusion of a number of different topics based on reviews of AETC reports, our review of the QOLS, meetings with leaders, MTI feedback on draft survey items, and the scientific literature. Like the trainee survey, the current MTI survey represents RAND's efforts to address the key goals of the survey system, while balancing survey length to avoid fatigue that could lead to MTIs not finishing the survey or rushing to finish it without necessarily reading carefully.

For example, as discussed previously, we considered assessing MTIs' self-reported engagement in misconduct, but on the basis of feedback from MTIs and AETC leadership, determined that it was unlikely that MTIs would feel sufficiently protected to reveal incidents of personal misconduct. To determine the prevalence of abuse and misconduct toward trainees, we determined that the best approach would be to use trainees' reports of victimization.

Similarly, we considered including items to address MTIs as victims of abuse and misconduct. This is an important topic that BMT may wish to consider in a separate future survey. However, we ultimately decided to exclude this from the current survey since the primary focus of the survey system was to detect abuse and misconduct toward trainees, and the repetition of similar questions with the different frames of references would considerably lengthen the survey.

There were additional topics that we considered for inclusion in the quality-of-life section of the survey, such as health and performance outcomes for MTIs. These items would be very useful for research purposes to try to link some of the stressors or environmental variables to actual outcomes. However, the purpose of the survey was to assess abuse and misconduct toward trainees as well as MTI work attitudes and perceptions of the environment. Thus, we decided that the survey space was better used to address potential antecedents of these outcomes identified in the scientific literature or raised by leaders, SMEs, and MTIs. AETC may instead wish to compare some of the survey findings with other data sources that do capture these types of outcomes.

Analyzing and Interpreting Results from the MTI Survey

Below, we provide an overview of recommendations for analyzing and interpreting results from the MTI survey. As one of the key deliverables for this study, RAND also developed a reporting template to help facilitate the below recommended analyses and tracking trends over time. This template is similar to the snapshot presented in Appendix D for the trainee survey.

Preliminary Data Validation

As recommended for the trainee survey, in order to ensure high confidence in the reported data, it is important to validate each MTI's data and to remove from the data set the responses of any MTI who did not complete the survey in good faith. First, at the end of the MTI survey, each MTI responds to the item: "How open and honest did you feel you could be when answering these survey questions?" The responses of any MTI who indicates that he or she was "not at all open or honest" should be noted and then dropped from the final data set, as these admittedly dishonest responses may contaminate the full data set. Like the trainee survey, we do suggest that the responses from this subset of participants be analyzed separately given the possibility that those individuals who had negative experiences may be likely to indicate they did not feel that they could be open and honest on the survey. Second, each of the five main abuse and misconduct sections (Sections VI–X) includes an item to assess whether MTIs are reading and responding to the questions carefully. For example, Section VIII includes the following item: "Please select 'Daily' for this item to help us confirm that MTIs are reading these items." An MTI who is responding quickly without reading items will not respond correctly to this item and can be identified as someone for whom confidence in the accuracy of his or her responses should be low. Five items in the questionnaire check if MTIs are reading and responding carefully (6.1e, 7.1k, 8.1n, 9.1k, 10.c). If a respondent answers any of these items incorrectly, he or she should be excluded from the final data set. Finally, we also recommend examining whether there was any variance in answers for the items in Sections I–IV. Because these sections contain some items that are reverse coded (i.e., items worded so that a highly positive response would be endorsed as "strongly disagree" compared with the other items in the scale, in which a positive response would be endorsed as "strongly agree"), there should be some variance in answers. Anyone who had the same answer through an entire section or no variance in his or her answers would not have been reading the questions carefully and should be excluded.

MTI Quality of Life

The majority of the constructs measured in the MTI quality-of-life part of the survey (Sections I–V) are designed as scales composed of multiple items. The average of the items should be taken to create a single score that represents the overall underlying construct (e.g., the average rating across the five items assessing commitment is used to represent overall commitment to the organization on a scale of one to five).[8] These constructs include organizational commitment, treatment of rookie MTIs, general MTI interpersonal treatment, leader treatment of MTIs, ethical leadership, and work-family conflict. When reporting the results from these scales, it will also be important to examine the variance in responses.

In other sections of the survey, we recommend that the items be analyzed and reported as a set of single items rather than as scales; this is done by examining the number and percentage of MTIs who endorse each response option. Specific sections with these items include job satisfaction, organizational support, perceptions of trainees, assignment and promotion opportunities, job stressors, work hours, sleep quality, sleep quantity, improvement of instructional skills, and participation in deliberate development courses.

---

[8] As a general rule, reliability will be higher when multiple items are used to measure an underlying construct rather than using single items to measure a construct. However, when multiple items designed to fit on a scale are not highly intercorrelated (i.e., internal consistency reliability), it suggests that multiple constructs or dimensions are being measured. Therefore, on future MTI surveys, an analyst should continue to reevaluate the internal consistency of scales on the MTI survey. As a rule of thumb, the internal consistency correlation coefficient should be above 0.70 (Cohen, 1977). A statistical technique known as factor analysis can also be used to evaluate whether the items should continue to be grouped together to form a scale.

Abuse and Misconduct Toward Trainees

Awareness of Bullying, Maltreatment and Maltraining, Unprofessional Relationships, Sexual Harassment, and Unwanted Sexual Experiences

For analyzing MTI awareness of abuse and misconduct, we recommend following the same procedures as described in Chapter Two for the trainee survey. However, it is important to note that the results from the MTI analysis should not be interpreted as estimates of prevalence. As described earlier in this chapter, results from these survey sections should be considered an assessment of the overall environment (MTIs' perceptions of occurrence of abuse and misconduct toward trainees). The extent of match or mismatch between MTI awareness of misconduct and trainee reports of misconduct can be considered an indirect measure of the success of the reporting systems.

MTI Perceptions of the Squadron Climate

Each abuse and misconduct domain also includes four items to measure MTI perceptions of the squadron climate. These items mirror the items on the trainee survey, and responses to the items should be averaged together to create a single scale score for each respondent. The average score across MTIs can then be reported to represent overall perceptions of the extent to which squadron leaders enforce policies and encourage the reporting of incidents, with higher scores indicating more-positive views.

MTI Reporting Norms

Each abuse and misconduct domain also includes four items to assess MTI perceptions of the general reporting environment or MTI norms for reporting behaviors. These items were not found to form a reliable scale, so they should be analyzed and reported separately by examining the number and percentage of MTIs who endorse each response option.

Clarity of Abuse and Misconduct Policies

Finally, Section XI of the survey includes items assessing the clarity of abuse and misconduct policies at BMT. These items are meant to be analyzed separately by examining the number and percentage of MTIs who endorse each response option.

Background Characteristics

For all of the scales and items in the MTI survey, we also recommend examining whether differences exist based on the length of time individuals have been MTIs and their primary duties. Examining potential differences through these background characteristics can provide leadership with greater insight about MTI well-being and the BMT work environment, as well as help leadership better address issues and tailor intervention efforts.

Summary

Thus, the MTI survey serves as a separate but complementary survey to the BMT trainee survey. It provides a framework for assessing both MTI quality of life and the extent to which MTIs are *aware* of trainees experiencing the same abuse and misconduct behaviors addressed by the trainee survey. The survey also includes a section on MTI perceptions of squadron climate related to each abuse and misconduct domain, perceptions of MTI reporting norms, and the extent to which related policies are clearly defined. Table 3.1 provides an overview of the final survey content.

Table 3.1. Overview of BMT MTI Final Survey Content

Survey Content	Number of Items
Background demographics	
Length of time as an MTI	1 item
Primary duty	1 item
MTI Quality of Life	
Section I: Job attitudes	
Organizational commitment	5-item scale
Job satisfaction	1 item
Section II: Work environment	
Organizational support	4 separate items
MTI interpersonal treatment	
Treatment of rookies	4-item subscale
Treatment in general	4-item subscale
MTI perceptions of trainees	6 separate items
Assignment and promotion opportunities	2 separate items
Section III: Leadership	
Leader treatment of MTIs	10-item scale
Leader ethical conduct	5-item scale
Section IV: Work and family stressors	
Work-family conflict	
Family interference with work	3-item subscale
Work interference with family	3-item subscale
Specific stressors	28 separate items
Average number of working hours	1 item
Average number of sleeping hours	1 item
Quality of sleep	1 item
Section V: MTI professional development	
Improvement of instructional skills	1 item
Deliberate development course participation	1 item
Awareness of Abuse and Misconduct Toward Trainees	
Section VI: Trainees bullying other trainees	
No subscales	6-item scale
Section VII: Maltreatment and maltraining by MTIs	
Maltraining	5-item subscale
Privacy violations	2-item subscale
Denial of services or rights	2-item subscale
Hostile comments	2-item subscale
Encouragement of mistreating other trainees	1-item subscale
Physical threats or force	5-item subscale
Section VIII: Unprofessional relationships with MTIs	
Attempts to establish a relationship	5-item subscale
Inappropriate exchanges of money	2-item subscale
Inappropriate social contact	3-item subscale

Survey Content	Number of Items
Relationship policy violations	6-item subscale
Section IX: Sexual harassment from anyone at BMT	
Sexist hostility	3-item subscale
Sexual hostility	4-item subscale
Sexual coercion	4-item subscale
Unwanted sexual attention	3-item subscale
Challenges to masculinity/femininity	2-item subscale
Section X: Unwanted sexual experiences committed by anyone at BMT	
Exposure of private areas of the body	1-item subscale
Unwanted sexual contact	1-item subscale
Attempted rape	1-item subscale
Completed rape	1-item subscale
MTI perceptions of squadron climate (repeated in each abuse and misconduct section)	
Bullying	4-item scale
Maltreatment and maltraining	4-item scale
Unprofessional relationships	4-item scale
Sexual harassment	4-item scale
Sexual assault	4-item scale
MTI reporting norms (repeated in each section involving potential MTI abuse and misconduct)	
Maltreatment and maltraining	4 separate items
Unprofessional relationships	4 separate items
Sexual harassment	4 separate items
Sexual assault	4 separate items
Section XI: Clarity of abuse and misconduct policies	
Bullying	1 item
Maltreatment and maltraining	1 item
Unprofessional relationships	1 item
Sexual harassment	1 item
Sexual assault	1 item
Closing questions	
Open and honest	1 item
Open-ended write-in response	1 item

4. Survey Participation and Administration

This chapter presents our recommendations regarding survey participation and administration for both the trainee and MTI surveys. These surveys should be administered via computer to permit necessary branching on survey questions (see Figure 2.1). Unlike Scantron or paper-and-pencil methods, computerized survey taking can also help to eliminate potential errors, save time, and conserve the resources required to conduct and analyze a survey. Below, we make recommendations regarding participant selection and guidance, survey timing, and promoting open and honest responses.

Participant Selection and Guidance

Trainees

Rather than conducting the survey with a random sample of trainees, **we recommend that *all* trainees take the survey**. Some abuse and misconduct behaviors may be relatively rare, so surveying all trainees will provide better prevalence estimates. In addition, surveying all trainees helps protect the identity of those who reveal abuse and misconduct. It can also help prevent offenders from targeting trainees whom they know will not be surveyed. Finally, because BMT involves a highly structured and monitored environment, it is much easier to build the survey administration into the current BMT schedule than to sample trainees for participation at random. This is similar to how end-of-course surveys for BMT are already conducted. All trainees would be required to attend a survey session with their flight, except in cases of emergency. We also recommend that all trainees be required to remain in the survey room for the entire survey session (45 minutes will allow time for instructions and the trainees who may need a bit longer than average to complete the survey). Permitting trainees to leave as they complete their surveys would provide incentive for them to decline to participate or rush through the survey, and could suggest to peers which trainees are indicating abuse and misconduct and thus are receiving more of the follow-up questions. For those same reasons, trainees should not be told that they must all remain until the last participating trainee completes the survey.

We also recommend that the survey be administered as an exit survey to all trainees who fail to complete BMT to determine if abuse or misconduct is a factor that predicts departure. The results from these surveys can be integrated into leadership reports but should also be analyzed separately to see if the responses differ from those of graduates.

Consent Options

Participation in the survey should be voluntary for all trainees. Given the sensitivity of the topics, it is important that no trainee feels forced to participate and potentially relive a traumatic event.

The survey presented in Appendix B includes our recommendations for the instructions and consent options provided to trainees. We recommend that trainees have the option to (1) participate in the survey; (2) decline to participate, but click through the survey so others do not know they opted out; or (3) decline to participate. Trainees also should have the option to skip any questions they do not feel comfortable answering or stop taking the survey at any time. Trainees who choose to not participate or who finish early should be allowed to study or rest until the session ends.

MTIs

We recommend that all MTIs at BMT who have been an MTI for at least one month be invited to participate in the survey. This survey is designed to collect feedback from MTIs on their work experiences and the BMT environment, and all MTIs who wish to participate should be allowed to do so. However, we recommend restricting the survey to MTIs with at least one month's experience so they have sufficient knowledge to answer the survey questions.

Consent Options

As with trainees, we recommend that all MTIs be required to attend a survey session, but actual participation should be voluntary. Requiring attendance helps ensure that supervisors or commanders do not prohibit MTIs from attending and that everyone has the opportunity to participate. The survey presented in Appendix C includes our recommendations for instructions and consent options. Like trainees, MTIs should be able to skip any questions they do not feel comfortable answering or stop taking the survey at any time. Individuals who decline to participate or finish early should be allowed to leave and resume their duties.

Survey Timing

Trainee Survey

We recommend that the survey be given to every class of trainees coming through BMT, with administration taking place as close to the end of BMT as possible. As part of developing the survey content and our proposed administration procedures, RAND conducted an analysis of data from different trainee week groups to see if the timing would affect responses (see Appendix A for a description of the test survey). We administered the survey to trainees who had completed weeks three or seven of BMT, as well as a small group of students who had just graduated from BMT and were starting their TT at Lackland Air Force Base in San Antonio, Texas. Overall, we found that the later the survey was administered, the more trainees were likely to report that they had experienced abuse and misconduct, which may reflect the greater window of opportunity for such incidents to have occurred.[9]

To further explore potential differences in the survey timing, the test also included a series of questions on how open and honest trainees would be on the survey at different points in time. We analyzed the results only for trainees who indicated on the test version of the survey that they were either somewhat or completely open and honest. Table 4.1 presents the results.

Table 4.1. Attitudes About Survey Timing Among Trainees Who Reported Being Somewhat or Completely Open and Honest on the Test Survey

Survey Timing	Very Comfortable Being Open and Honest (%)	Somewhat Comfortable and Uncomfortable (%)	Very Uncomfortable (%)
Before BMT graduation (N = 980)	77 (74, 80)	20 (17, 23)	3 (2, 4)
After BMT graduation and during TT (N = 972)	87 (85, 89)	11 (9, 13)	2 (1, 3)
After graduation from TT (N = 967)	85 (83, 87)	12 (10, 14)	3 (2, 4)

NOTE: 95-percent confidence interval lower and upper bounds reported in parentheses.

As the table shows, the vast majority of participants would feel comfortable under all three administrative options. However, taking the survey after graduating from BMT would be ideal.

Based on these findings, we explored potential administrative options in meetings with AETC headquarters and BMT leadership, particularly the administration of the survey in TT. However, given that TT takes place at multiple locations across the country with variations in the timing that students attend and the facilities available for administration, we ultimately concluded that the difficulties in coordinating standardized data collection would outweigh the benefits. Furthermore, a majority of trainees indicated they would be comfortable taking the survey during BMT.

MTI Survey

Assuming that no other routine MTI surveys are introduced, we recommend conducting this survey every six months to a year, with more-frequent surveys following spikes in abuse and misconduct incidents, major personnel turnover, and changes to policies or programs. During our test of the MTI survey, we asked how often the survey should be

---

[9] Some behaviors, such as unwanted sexual experiences and unprofessional relationships, elicited little variability in responses, so we were not able to fully examine week group differences for all types of abuse and misconduct.

conducted so that MTIs can provide leadership with feedback. As Table 4.2 shows, the majority of participants recommended six months to a year. Many MTIs who provided an "other" response indicated that it depended on whether leadership actually took actions based on the survey.

Table 4.2. MTIs' Recommendations for Survey Frequency

Survey Frequency	Percentage of MTIs (N = 224)
Three months	13 (*10, 16*)
Six months	34 (*30, 38*)
One year	37 (*33, 41*)
Other	16 (*13, 19*)

NOTE: 95-percent confidence interval lower and upper bounds reported in parentheses.

Six months to a year also helps address the desire among leadership for problems to be detected quickly and for the MTI survey to closely track the trainee surveys, which are weekly. If there are spikes in abuse or misconduct or a lot of changes taking place in the BMT environment, conducting the survey at six-month intervals may be helpful; otherwise, once a year should be sufficient. This recommendation also assumes that that no other routine and potentially competing MTI surveys are introduced.

To increase participation rates, we recommend that BMT hold several different survey sessions for MTIs, scheduled around different shifts and over multiple days so that MTIs can select the session that will best work with their schedules. The survey and times should be advertised to MTIs at least two weeks in advance. (Sample content for a survey recruitment letter is provided in Appendix E.) Finally, squadron commanders should be accountable for removing barriers to MTIs attending the sessions and for providing them time during their normal duty hours to complete the survey.

Promoting Open and Honest Responses

Having a survey that will allow individuals to feel that they can be as open and honest as possible is essential given the sensitivity of the topics. Therefore, **we recommend that the trainee and MTI surveys be anonymous or confidential to the greatest extent possible**. Although having identifiable surveys can permit a direct investigation into any reported incidents and allow leadership to contact someone with follow-up questions, this loss of anonymity will likely reduce both participation and the accuracy of the data collected. For example, trainees who are afraid to make an official report of abuse or misconduct would be unlikely to be honest on an identifiable survey. Similarly, MTIs who are unwilling to raise a concern directly with someone in their chain of command would be unlikely to participate in or be honest on a survey that identifies them. Instead, having a survey system that is as anonymous as possible provides a reporting channel that allows participants to be open and honest without fear of retribution.

For the test of the survey instruments, RAND took several steps to protect the anonymity and confidentiality of trainees and MTIs. First, for the trainee survey, no MTI or other Air Force official was permitted to enter the survey room during the administration period. Similarly, no Air Force officials or members of BMT leadership were permitted in the MTI survey session. Second, with the support of Lackland Air Force Base computer specialists, a local computer network was configured to work without the normal common access card (CAC), so trainees and MTIs did not have to log on in a way that would identify them. Third, demographic questions were limited to the bare minimum to reduce the risk of identification by inference. Finally, all computer screens were outfitted with privacy protectors, which prevent individuals seated at neighboring monitors from viewing the content on other screens. To assess the effectiveness of these procedures, both the trainee and MTI surveys asked participants how open and honest they could be when answering the survey questions (conducted under the above-described conditions). Of those who participated, only three of the 1,004 trainees and one of the 280 MTIs indicated they could not be open and honest.

To assess how trainees and MTIs might respond to alternative methods, the test surveys included items to assess how comfortable trainees and MTIs would feel being open and honest under various circumstances in the future. The level of comfort may not preclude respondents from being open and honest on a survey, but it gives an indication of the methods that are most likely to result in open and honest answers. As Tables 4.3 and 4.4 show, many trainees and MTIs would be less comfortable answering the same questions if their survey participation were more identifiable. Among trainees who were either somewhat or completely open and honest on the test survey, 38 percent would be very uncomfortable if they had to use their CAC card to take the survey, even if the Air Force promised not to link the survey to them, and 45 percent would be very uncomfortable if they had to enter their names on the survey. More than 40 percent of MTIs who were somewhat or completely open and honest on the test survey would be very uncomfortable being open and honest if survey administration required a CAC card or name on the survey. The survey administrator is also important; participants were less comfortable with military personnel administering the survey compared with civilians inside or outside the Air Force. Finally, more trainees would be comfortable if their results were reported after they graduated from BMT, and more MTIs would be comfortable if the survey was conducted after they completed their tours.

Table 4.3. Attitudes About Alternative Survey Administration Conditions Among Trainees Who Reported Being Somewhat or Completely Open and Honest on the Test Survey

Survey Item	Very Comfortable Being Open and Honest (%)	Somewhat Comfortable and Uncomfortable (%)	Very Uncomfortable (%)
Your CAC card must be in the computer while you take the survey, but the Air Force promises not to use it to link your survey to you (N = 949)	28 (25, 31)	33 (30, 36)	38 (35, 41)
You must enter your name on your survey (N = 949)	25 (22, 28)	30 (27, 33)	45 (42, 48)
The survey is conducted by civilian analysts who work for the Air Force (N = 962)	78 (75, 81)	20 (17, 23)	3 (2, 4)
The survey is conducted by analysts who are Air Force military personnel (N = 960)	67 (64, 70)	25 (22, 28)	8 (6, 10)
The survey is conducted by analysts outside of the Air Force (N = 959)	78 (75, 81)	18 (16, 20)	5 (4, 6)
The overall survey results for your week group are reported to command while you are still at BMT (N = 943)	44 (41, 47)	32 (29, 35)	24 (21, 27)
The overall survey results for your week group are reported after you have graduated from BMT (N = 930)	64 (61, 67)	30 (27, 33)	6 (4, 8)

NOTE: 95-percent confidence interval lower and upper bounds reported in parentheses.

Table 4.4. Attitudes About Alternative Survey Administration Conditions Among MTIs Who Reported Being Somewhat or Completely Open and Honest on the Test Survey

Survey Item	Very Comfortable Being Open and Honest (%)	Somewhat Comfortable and Uncomfortable (%)	Very Uncomfortable (%)
Your CAC card must be in the computer while you take the survey, but the Air Force promises not to use it to link your survey to you (N = 226)	35 (31, 39)	22 (18, 26)	43 (39, 47)
You must enter your name on your survey (N = 229)	36 (32, 40)	21 (17, 25)	42 (38, 46)
The survey is conducted by civilian analysts who work for the Air Force (N = 228)	64 (60, 68)	29 (25, 33)	7 (5, 9)
The survey is conducted by analysts who are Air Force military personnel (N = 227)	60 (36, 64)	26 (22, 30)	14 (11, 17)
The survey is conducted by analysts outside of the Air Force (N = 227)	77 (73, 41)	19 (16, 22)	4 (2, 6)
The survey is conducted after you have completed your tour as an MTI (N = 228)	67 (63, 71)	25 (21, 29)	8 (6, 10)
The survey is conducted while you are still an MTI (N = 229)	59 (55, 63)	36 (32, 40)	5 (3, 7)

NOTE: 95-percent confidence interval lower and upper bounds reported in parentheses.

Recommended Actions

To promote open and honest responses, we recommend the following steps. First, the surveys should not require participants to provide identifying information, such as names and ID number. Second, although computer use in the Air Force often requires the use of a CAC (which identifies the individual using the computer), AETC has spearheaded an effort to use a token system that will provide anonymous computer access for taking the surveys. Third, we recommend the use of privacy screens or dividers on all the computers to prevent participants from seeing each other's responses. Fourth, we recommend that data not be analyzed until after trainees have graduated from BMT. For practical reasons and to obtain systematic feedback on a regular basis from MTIs throughout their lengthy assignments, AETC should not wait to administer and analyze MTI data after MTIs have completed their duty tour. We do recommend that MTIs be provided with an opportunity to provide additional feedback on their experiences and the issues addressed in the survey once they have completed their tours though. Fifth, we recommend including explicit language in the instructions and consent statement at the start of the surveys that clearly states the safeguards taken to protect participants' identities and the benefit of their participation in the survey to BMT and future trainees.

Finally, we recommend having civilian analysts who report to an office in AETC headquarters, not to BMT leadership, oversee survey administration. If the individuals administering the survey are in their chain of command or directly linked to BMT leadership, participants may worry that they are being watched and feel pressured to provide certain responses. Placing the analyst in the chain of command under the BMT leadership could create a real or perceived conflict of interest that could undermine the credibility of the surveys and influence the way that the survey results are analyzed and presented to AETC headquarters. Although having an independent analyst outside the Air Force may be ideal, we realize that this may not be practical for continuous survey administration.

For similar reasons, MTIs should not be allowed to remain in the room, interrupt, or have an opportunity to observe a survey session in any manner. The same process should apply to MTI survey sessions: BMT leaders should not be allowed to attend, interrupt, or monitor survey sessions. Trainees and MTIs must feel that their participation is truly voluntary and that their responses are confidential to answer the survey honestly.

Although an anonymous or confidential survey does not facilitate direct investigation into any incidents, it can inform action in the following ways:

- Advise leadership about the attitudes and behaviors of the members.
- Help identify spikes in certain types of abuse and misconduct that should prompt follow-up discussions with trainees or MTIs to learn more.
- Allow for a comparison between the survey results and known incidents to assess how many individuals do not feel comfortable coming forward to make a report.
- Provide information that is not available through current formal reporting channels, including barriers to reporting that may exist, so that command can address the issue.

It is important to note that law enforcement or BMT leaders should never try to deduce who filled out any particular survey. This would severely undermine the intent of the survey and lead to less open and honest answers in the future.

5. Reporting Results and Taking Action

Organizational surveys can be an important tool for measuring the prevalence of behaviors such as abuse and misconduct, job attitudes, organizational strengths, and opportunities for improvement. However, assessment is only the first step toward making improvements within an organization. To maximize the value of the surveys, we recommend that AETC follow up with (1) analyses and trend tracking over time, (2) triangulation with other relevant data sources and follow-up data collection to better understand the results, (3) a systematic process for reporting results to senior Air Force leaders and other key stakeholders, (4) prioritization of problem areas and setting goals for improvement, and (5) implementation of new policies and improvement plans.

Analyses and Tracking Trends

The trainee and MTI surveys are designed to provide key data and feedback to help guide the interventions and policy changes needed to make improvements at BMT. To ensure that results are interpreted accurately, it is strongly recommended that a qualified analyst with a background in social sciences and statistics be assigned to conduct the analyses and interpret the results.

First, the analyst should examine the survey data to ensure that it meets scientific standards for analysis and interpretation. For example, the sample sizes required for the different types of analyses and comparisons being requested should be carefully examined to confirm that there is adequate power to detect a significant effect or association. Analyses that lack the appropriate sample size may lead to misinterpretation, such as overstating the size of effects that are found to be statistically significant or thinking that not finding a difference between groups is evidence that no difference exists. In contrast, very large sample sizes may result in statistically significant findings even when differences are very small. In this case, additional steps should be taken to ensure that statistically significant findings are meaningful. Determining meaningful differences in large sample sizes requires more judgment and will depend to some extent on the perspective of the person evaluating the differences. It is important for the analyst to guide senior leaders in determining which results are meaningful and should be given attention.

Once the data have met minimum requirements for analysis, a thorough review and analysis is needed to identify relationships between measures and over time. By examining trends in the survey data over time, the surveys can operate as a monitoring system for the BMT environment that not only benchmarks abuse and misconduct but also measures overall MTI quality of life. Survey results over time can also be used to determine the effectiveness of different interventions (e.g., changes in policy or training programs), or results can be examined in relation to retention rates and other important organizational outcomes. For example, as changes in staffing and policies lessen the MTI workload, the survey can provide insight into whether related stressors are decreasing for MTIs. For both of these objectives, though, it is strongly recommended that leadership consult with the analyst to ensure that these types of analyses would meet scientific standards. For example, analyses to evaluate the effectiveness of an intervention may not be appropriate if multiple interventions or changes were made at the same time across BMT, as it would be impossible to determine which intervention or combination of interventions led to improved outcomes. Similarly, when linking survey results to other criteria, it is important to evaluate the quality of the measures (e.g., reliability) for these other criteria.

The steps needed to make improvements at BMT will also heavily depend on the results of the analyses and the level of confidence that observed differences or trends are significant. Minor variations or fluctuations in average scores on survey measures over time are common and should be expected. And it is critical to take into account the sample size when examining trends over time given that some abuse and misconduct behaviors are relatively rare. What could look to be a large percentage change may actually only be the difference of a few people or incidents. Therefore, it is important to carefully review the statistical and practical significance of the results before reallocating time and resources to a new or emerging problem. Although statistical analyses can help to identify and describe patterns in the data, additional steps are often needed to understand why these patterns exist.

Triangulation with Other Data Sources and Additional Data Collection

The survey data should not stand in isolation from other indicators of abuse and misconduct at BMT or data on conditions that increase risk of incidents. Triangulation with other data can help in constructing an integrated feedback system. These include:

- trainee data—such as BMT mental health screenings, end-of-course surveys, and data from trainees' comment sheets
- MTI data—the MTI quality-of-life survey, data to screen MTIs, personal information files, the MTI end-of-course survey, and manning data
- general-population surveys—including Air Force-wide assessments reported by unit, installation, or major command
- official incident data
- data from hotlines, SARCs, chaplain metrics, and BMT production data, such as injury and graduation rates
- security camera surveillance footage.

For example, data from BMT's official channels for reporting abuse and misconduct could be examined along with the survey results to explore the proportion of incidents that are unreported. Similarly, information collected through anonymous trainee comment sheets and the BMT hotline may provide insight into areas of concern that arise on the survey. Taken together, all these sources help form a more integrated feedback system that can help BMT and Air Force leadership better prevent and respond to abuse and misconduct. A more detailed description of these additional data sources can be found in Appendix F.

In some cases, additional data collection may also be needed to better understand the survey results. In these cases, we recommend that leadership consider conducting small focus groups or interviews. For example, a SARC could hold focus groups with trainees about the barriers to reporting, or interviews may be useful for understanding attitudes about interpersonal treatment among MTIs. These follow-up interviews, focus groups, or some other form of qualitative data collection can be critical for leadership to fully understand certain results and how to address them. Follow-up data collection can also help explain trends over time, differences among groups, or why a new problem has emerged. In addition to helping to better diagnose problems, interviews and focus groups can also suggest why certain policies or changes have been effective. This advantage may be particularly relevant when several initiatives have been implemented at the same time.

It is important to note that at no time should the follow-up data collection involve any type of investigative process. Any attempt to identify who responded in a certain manner on the surveys or what happened in a particular incident will undermine the intent of the surveys to serve as a confidential feedback channel.

Systematic Process for Reporting Results

A critical step in developing an effective feedback system is to have a systematic process for reporting results. This includes the frequency with which results are shared and how and to whom the results are distributed.

Given the different types of information and frequency with which we recommend administering the trainee and MTI surveys, we have different recommendations for how often results from these surveys should be shared with leaders. In the case of the trainee survey, we recommend that results be reported quarterly. Although data collection will occur weekly, it is important to have a large-enough sample to provide more-stable results and to examine potential differences in responses. Quarterly reports will ensure a sufficient sample size while still providing a frequent check on potential abuse and misconduct occurring in BMT. The data analyst can also check more frequently for substantial spikes in abuse and misconduct to ensure that nothing is being overlooked. For the MTI survey, we recommend that data be collected every six months to a year, so results should be reported in coordination with the data collection cycle.

For both the trainee and MTI surveys, we recommend that reports of the results be provided and tailored to each leadership level (i.e., wing, group, and squadron) as well as to the AETC Recruiting, Education and Training Oversight Council (RETOC) and other relevant stakeholders (e.g., SARCs, chaplains). Sharing the results broadly can help facilitate a dialogue and build commitment for taking action across leadership levels and with stakeholders on how to best address any issues that arise. As mentioned, the results should be presented along with data from other relevant sources and include information about any actions taken to address key issues. Given the different roles at BMT, the focus and level of detail should also be carefully tailored to each group's needs so that it can focus on the most relevant information. It is critical for leadership to identify who is responsible for doing what with the results or whether certain results are provided for information only.

Prioritizing Problem Areas and Setting Goals

Attempting to address too many areas at the same time increases the likelihood of diluted resources and ineffective change. Therefore, leadership should consider developing criteria for prioritizing the problem areas. The criteria may include the severity of the problem, cost, available resources and expertise, and likelihood that an intervention will have an impact.

Feedback received by relevant stakeholders—squadron leaders, other Air Force leaders, SARCs, chaplains, security personnel, and legal personnel—can also help with prioritization. Identifying representatives from each of these stakeholder groups to work as a team can make the process more effective. For example, representatives can help organize the feedback and concerns of stakeholder groups as changes are planned and implemented. Representatives should have good working relationships across organizational levels and between units, and have the respect and trust of the stakeholder groups they represent. Further, representatives should be highly committed to the Air Force and BMT's long-term success.

We also encourage leadership to involve MTIs in feedback sessions to discuss and better understand the survey results and help prioritize problem areas before making changes. This will help build commitment and readiness to change and increase the probability of progress. Participation can also help MTIs understand why change is needed and what possible benefits are associated with the change. However, it is important to note that participation in feedback sessions can result in disengagement and apathy toward future surveys if goals are not established and actions implemented to reach them. Therefore, one of the most important steps is collaborating with MTIs to set clear and specific goals for improvement. These goals should be documented and shared broadly among MTIs and with AETC headquarters and BMT leaders to promote awareness of the problems and the steps BMT is taking to solve them. Sharing goals can also clarify the roles and responsibilities that all relevant parties have for supporting change efforts.

Implementation of New Policies and Improvement Plans

Taking action and following through are key to effecting change. Reforms should be carefully planned and implemented to minimize the perception that policies are constantly shifting, and leadership should give great consideration to the timing and frequency of changes made within BMT. Once changes have been implemented, it is important that leadership and representatives from the different stakeholder groups communicate regularly with those affected to ensure that the reforms are being accepted and followed—and are working as intended.

Reinforcing realistic timelines and goals is another important step. It may take several months or more before anticipated outcomes are realized. It is important to maintain a supportive environment that encourages communication and feedback to support a high level of commitment to new programs, policies, and other changes.

An environment that facilitates communication will help leadership address any problems or resistance that may surface. Regular progress reports should also be developed and shared with those affected by the change to show which actions have been taken, which actions are being planned, and which objectives have been met. These reports should build not only commitment to the survey system but also trust in AETC and BMT leadership.

6. Conclusion and Additional Recommendations for BMT

The Survey System as a Leadership Feedback Channel

This report describes the integrated survey system RAND developed for the Air Force to implement at BMT in response to incidents of abuse and sexual misconduct. The trainee and MTI surveys—designed for ongoing, routine administration—are intended to help AETC track trends and detect problems before they proliferate. The results of these complementary surveys should be integrated into an overall feedback system to give leadership a more complete picture of "ground truth" and point toward corrective actions.

The survey system is designed to augment rather than substitute for direct interaction and monitoring by leadership and by professionals, such as chaplains, medical personnel, and SARCs. The surveys are a more private alternate to the multiple channels that already exist for trainees and MTIs who are comfortable directly reporting problems and being identified with that report. The problems that led to the crisis in 2012 were able to escalate in part because individuals were afraid to come forward. The survey system's value will be undermined if participants are required to enter any identifying information to access the survey; if the survey is administered by members in the chain of command or anyone being evaluated on the surveys; or if commanders, law enforcement, or legal staff attempt to discern a participant's identity based on survey responses, regardless of their good intentions.

The trainee survey focuses on the abuse and misconduct that trainees may have experienced or witnessed, their decisions and experiences related to reporting abuse or misconduct, perceptions of the squadron climate, and perceptions of BMT systems designed to facilitate feedback to leadership and support trainees. The trainee survey should be administered at the end of BMT to capture as much of BMT as possible and to avoid a postsurvey window that potential perpetrators could exploit. Although survey participation should be voluntary, all trainees should be required to attend the survey sessions so that MTIs cannot pull out trainees who might report them or target groups that will not be surveyed. Results should be analyzed quarterly to provide leadership with timely feedback, but the weekly results for the most serious misconduct should be scanned by an analyst to quickly identify and alert leaders to any new developments

The MTI survey focuses on MTI quality of life and awareness of abuse and misconduct toward trainees. Assuming that no other routine MTI surveys are introduced, we recommend conducting this survey every six months to a year, depending on the survey results and changes being made at BMT. For example, the survey should be given more frequently to follow up on periods of great stress; major changes to programs, policies, or working conditions; or MTI complaints or complaints against MTIs. If results remain positive and stable for a period of time,

AETC could administer the MTI survey annually instead and rely on the trainee survey to identify new problems.

The survey results should be analyzed and interpreted by a qualified analyst with a background in the social sciences and statistics. So that change can be assessed over time, most survey items should remain consistent. However, we recommend continued refinement and validation of the measures beyond what was possible to conduct in the scope of this project. The length of the survey should also remain roughly the same. The questions are not intended to be comprehensive, and AETC needs to be mindful of survey length both to avoid survey fatigue among participants and to ensure that analyzing and preparing the results remain manageable. However, the survey should be updated to reflect changes in terms, programs, and policies (e.g., the name of the SARC position is changed, a new reporting channel is introduced). We would also advise against collecting additional demographic data to reduce both the perception and the reality that someone might try to use the survey results to identify individual respondents. We also recommend retaining abuse and misconduct items even if they are never or rarely reported: This is a sign that BMT is working as it should, and removing avenues for detecting problems could invite them to return.

Finally, this survey system could be adapted for use in other Air Force settings. Potential contexts for adaptation include Technical Training, officer entry-level training, and Air Force–wide surveys. The survey could also be adapted to training in other branches of the military, but would need considerable review to match their training policies and practices.

Additional Reinforcement of the Leadership Feedback System

General Edward A. Rice Jr., the sponsor of this project while he was the AETC commander, asked RAND to consider the overall leadership feedback system that would enable AETC to address abuse and misconduct and identify gaps. Many changes were recommended and adopted following AETC's own investigations and evaluation of its system. We offer several additional observations for consideration.

Routinely Monitor Security Camera Footage

AETC has installed additional security cameras throughout the BMT area to help leadership combat forms of abuse and misconduct that could be captured visually, but footage is only reviewed if a complaint has been registered. If manpower permits, those recordings could be regularly monitored so that leaders could be alerted in a timely manner to suspicious behavior that might be going unreported. This would provide another means to detect incidents that might not otherwise be reported. In addition to abuse and misconduct, this footage may reveal other prohibited behavior, such as theft.

Evaluate Training That Prepares Trainees to Identify and Report Abuse and Misconduct

Throughout BMT, trainees are taught what behavior is and is not acceptable in the Air Force and the possible courses of redress should they witness or fall victim to such behavior. This is the process through which BMT teaches trainees about their expected roles and responsibilities within the leadership feedback system. That training and education have been substantially revised since 2012. We recommend evaluating the training to assess whether trainees comprehend and apply the material as AETC intends—that is, are they able to sufficiently identify what behavior they should be reporting and the various reporting channels they can use to alert leadership to problems or to seek help? According to well-established evaluation methods, the first step would be establishing the program outcomes to be evaluated. Table 6.1 presents some key outcomes a training evaluation could measure (Kirkpatrick and Kirkpatrick, 2006), as well as some examples of those outcomes. Trainees should have a good understanding of which MTI training techniques the Air Force considers appropriate and which are considered maltreatment or maltraining and should be reported. Trainees should also understand Air Force definitions of sexual harassment and sexual assault so they can conduct themselves accordingly and can respond appropriately if they witness or experience this behavior.

It would be valuable to confirm through a systematic formal evaluation that the training is meeting its goals in preparing trainees to identify and to be able to report abuse and misconduct to BMT leaders. Given the amount of material trainees are expected to learn while at BMT and that many may find the training conditions to be stressful, it is possible that these lessons are not fully or correctly absorbed. The training evaluation would only be needed every few years unless the training changes, the number of incidents increases, or reporting is lacking. Evaluation items should not be added to the surveys RAND developed because the surveys would grow too long, and weekly evaluation is unnecessarily fine-grained for an unchanging training. Whenever feasible, scientific training evaluation designs, including pre- and post-training assessments and random assignment to pretraining control groups, can further strengthen any education and training programs the Air Force develops.

Table 6.1. Primary Training Outcomes to Evaluate

Key Outcomes	Examples
Instruction: Was the instruction delivered as intended?	• Training/qualifications of instructor • Instruction accurately conveys policy, leadership expectations, and other information • Achieved intended depth of material and time allotment • Instructor ability to respond to questions
Reactions: Were participants satisfied with the program?	• Reactions to the program content • Reactions to the instructors
Learning: What did participants learn in the program? Did they learn anything new? Did they learn what the training was meant to teach them? Can they apply that knowledge to novel scenarios?	• Knowledge of acceptable and unacceptable MTI training practices • Increase in sexual assault–related knowledge (e.g., definitions and statistics on sexual assault) • Increase in knowledge and skills for how to recognize and react to risky situations • Decrease in rape-supportive attitudes (e.g., gender-role stereotypes) • Knowledge of reporting channels
Behavior: Did the participants change their behavior based on what was learned in the program?	• Increase in bystander intervention behaviors • Self-reported decrease in behaviors likely to increase risk of sexual assault
Results: Did changes in behavior produce desired results? Did the program positively affect the organization?	• Decrease in trainee-initiated abuse and misconduct • Increase in reporting of abuse and misconduct

Follow Up with Victims and Witnesses Who Filed Reports of Sexual Assault

Feedback on the experiences of trainees who have reported sexual assault or other serious complaints of abuse and misconduct can help AETC leadership identify negative experiences that violate standards or policy and could deter other victims and witnesses from coming forward. Examples of negative experiences would include the behavior continuing despite it having been reported, MTI or peer retribution for reporting, inappropriate comments or questions from leaders or investigators, and violations of confidentiality by those authorized to offer it. Of course, those negative experiences might make some victims and witnesses reluctant to engage in any further conversation with AETC leadership about the behavior or the process.

The survey system is not the right vehicle, however, for collecting feedback from this population. First, the number of reports will likely be too few each cycle or even each quarter to warrant inclusion in the survey. With such small numbers, it would not be possible to analyze the survey data statistically. Second, other survey participants might be able to identify abuse victims if their surveys took longer to complete because of follow-up questions. Furthermore, some of these processes will extend beyond BMT, after the survey had been administered.

Instead, we recommend that the Air Force explore other feedback mechanisms for monitoring how well AETC is managing victim care and responding to incident reports. Domains should include victim experiences with and perceptions of the reporting process, the judicial system, the victim care and advocacy system, and unit leadership reactions to reports. Someone outside these systems, such as a SARC or victim advocate from another major command, could be asked to reach out to victims and witnesses who have filed reports of incidents and ask that they volunteer to discuss their experiences. Note, we are not advocating trying to use the BMT survey to identify and follow up with trainees who indicated they had an unwanted sexual experience. As trainees move into TT or their first duty stations, the assessment could also address the continuity of support and care for victims across these transitions.

Feedback efforts will have to be developed with great care to ensure that they are not intrusive to victims, and that access to information about the identities of those who filed reports continues to be severely restricted. Because of the sensitive nature of this inquiry for victims in particular, it should be conducted through the communication mode that is both most feasible and comfortable for the victim, and the individual conducting the interviews should be equipped to address victim distress. As always, victims' privacy should be protected: The findings may not be anonymous, but they should be distributed to as few recipients as necessary to address any issues raised. SARCs already often follow up with victims as a part of care coordination: A new feedback system could at a minimum request aggregate, standardized updates from SARCs similar to the aggregate, standardized reports that SARCs prepare about initial reports of sexual assault.

Create an Online Central Repository Accessible to Key Leadership and Support Professionals Only

Our final recommendation to address a gap in the leadership feedback system is to create an online central repository for sharing among BMT leaders the various types of feedback that can serve as indicators of abuse and misconduct. In addition to the survey results, there are other important sources of data on the BMT environment and abuse and misconduct, such as end-of-course surveys and official incident data. Any information gathered will have the greatest value if analysts and leaders can review it in conjunction with other BMT data sources and if it is accessible to the different levels of leadership and support professionals who could use it to address abuse and misconduct. As we met with groups of SMEs to learn about these data sources, we also learned that they were not always aware of each other's data, when new data were available, and how they could be accessed. This limits the ability of leaders within BMT to connect the dots across these sources of information and to identify vulnerabilities in the system and those who may be exploiting them. Therefore, having a central repository to house these data can help facilitate data triangulation and sharing.

Access to this repository should be limited only to those with a legitimate use for it, such as the members of the AETC RETOC and the AETC Community Action Information Board and Integrated Delivery System, which are already charged with facilitating oversight and information sharing. As an archive, this repository would also protect against loss of historical trend data should the individuals who collect and analyze the data or write the reports leave AETC or their computer files become damaged (note that this refers to aggregate-level data, not individual-level data files, which should remain confidential and under more restricted use). AETC headquarters could require that all briefing slides, reports, executive summaries, talking points, and other forms of data reporting be posted in an organized fashion to that repository, and the website could be configured so that leaders who are members receive an alert when new information has been posted.

AETC has made great strides toward increasing its monitoring of the BMT environment and implementing reforms to improve its ability to dissuade, deter, detect, and hold accountable those responsible for abuse and misconduct. The survey system developed by RAND provides a way for trainees and MTIs to report abuse and misconduct toward trainees anonymously or at least confidentially and without fear of embarrassment or reprisal. It makes a unique contribution to the leadership feedback system. By institutionalizing this survey, AETC has ensured that leaders will be alerted in a timely manner to abuse and misconduct long after the subject has disappeared from the headlines.

Appendix A. Methodological Details on the Development of the Survey Content

This appendix describes the process RAND undertook to develop the trainee and MTI surveys. To the extent possible, the research team drew on established measures in the scientific literature with strong psychometric properties (i.e., the scale consistently and appropriately measures the construct it intends to measure) for inclusion in the surveys. In some cases, though, existing measures are not well suited to the BMT context or the constructs being measured are BMT specific and there are no existing measures (e.g., maltreatment and maltraining, unprofessional relationships). In these instances, we developed new items and scales specifically for these surveys, which were put through multiple reviews by SMEs and other key AETC stakeholders, as well as an initial test to help examine item and scale performance. It is important to note, however, that given the project time frame and resource constraints, it was not feasible to complete all traditionally recommended steps for development of these new scales in the scientific literature (e.g., Hinkin, 1998). Therefore, we recommend continued refinement and validation of the items and scales in these surveys to ensure the best possible survey system for monitoring abuse and misconduct in the BMT environment.

The overall development of the surveys proceeded through the following steps, which are described in detail in the sections below:

- Step 1: Selection of survey content domains
- Step 2: Selection and development of survey measures
- Step 3: Review and refinement of draft surveys
- Step 4: Test of the surveys.

Step 1: Selection of Survey Content Domains

The first step in the survey development process was to select the survey content domains for both the trainee and MTI surveys. To better understand AETC's goals for developing an integrated survey system, RAND first met with AETC headquarters, BMT leaders, and other relevant SMEs and stakeholders at BMT. The meetings continued throughout the surveys' development. After the initial discussions, RAND reviewed the available AETC reports describing the recent incidents of abuse and misconduct, as well as other relevant DoD and AETC materials and the scientific literature. These included:

- reports on the commander-directed investigation of the 2012 BMT incidents, including data from focus groups and a survey of trainees
- briefings documenting AETC plans and actions for restoring the trust at BMT

- DoD, Air Force, and AETC laws and policies (e.g., BMT Rules of Conduct, Administration of Military Standards and Discipline Training, Duty to Report, Professional and Unprofessional Relationships, Sexual Assault Prevention and Response, UCMJ)
- relevant trainee course material (e.g., sexual assault prevention and response, human relations, trainee rights and duties, preventing forbidden relationships)
- MTI course material on how to train
- a review of other data sources and feedback collected at BMT (e.g., trainee end-of-course surveys, MTI QOLS, occupational analysis data on MTI job requirements, trainee comment forms)
- relevant DoD and Air Force surveys and reports on sexual harassment and assault (e.g., DMDC Workplace and Gender Relations Surveys, DoD Annual Reports on Sexual Assault, Report on Sexual Misconduct Allegations at the U.S. Air Force Academy)
- scientific literature on sexual harassment, sexual assault, and workplace aggression and misconduct.

Based on our meetings and review of relevant materials, we then outlined a general plan for the surveys, with the goals of detecting incidents of abuse and misconduct in the training environment and providing data to help leaders understand what actions to take to reduce abuse and misconduct. We identified the following five core domains of abuse and misconduct to assess through the surveys:

- trainee bullying
- maltreatment and maltraining by MTIs
- unprofessional relationships with MTIs
- sexual harassment from anyone
- unwanted sexual experiences committed by anyone.

Step 2: Selection and Development of Survey Measures

In selecting and developing appropriate measures for our trainee and MTI surveys, we drew extensively on the materials we reviewed in step 1 and on established survey measures in the scientific literature. In some cases, we found previously validated measures we could use or adapt to the BMT context. In other cases, we developed new measures specifically for our surveys. The sections below provide an overview of the selection and development of the different measures.

Trainee Survey

Section I: Trainees Bullying Other Trainees

A variety of scales have been developed to measure bullying among children and adolescents (Hamburger, Basile, and Vivolo, 2011) and among adults in the workplace (Cowie et al., 2002). However, we found that many of these established scales were not fully appropriate for BMT, because they focus on populations that are too young (elementary and middle school) or include behaviors that are not fully applicable (bullying in an office setting). Additionally, many measures were copyrighted for use. As a result, we chose to develop a measure of bullying specifically tailored to the BMT context. We drew on established scales for bullying victimization outlined in a recent compendium of bullying measures by the U.S. Centers for Disease Control (CDC) (Hamburger, Basile, and Vivolo, 2011) that had good psychometric properties, were short in length, and were relevant to the age group. To be consistent with best practices, we used behaviorally specific items designed to capture the three dimensions of bullying most established in the literature: verbal, physical, and social exclusion/manipulation. We also included additional items that may be particularly relevant to the BMT context, such as trying to get another trainee in trouble with an MTI, for a total of nine draft items to measure bullying. We then further refined and reduced these items in steps 3 and 4 of our survey development process, resulting in a total of six items for inclusion in the final survey.

Section II: MTI Maltreatment and Maltraining of Trainees

Maltreatment and *maltraining* are terms specifically used by the Air Force to refer to a wide range of behaviors violating approved training methods and appropriate interactions between MTIs and trainees. Therefore, there are no established measures to incorporate or adapt for use on the surveys. Instead, using the Air Force definitions as a foundation, we searched the research literature for related constructs to identify additional content areas for item development. Search terms included combinations of the following: *verbal and physical aggression, abusive supervision/leadership, discrimination, harassment, workplace violence, counterproductive work behaviors, organizational deviance, misconduct, incivility, mistreatment, hazing,* and *emotional abuse*. This search yielded a number of scales that we considered adapting to a military context and BMT; however, most scales included behaviors that were beyond the scope of this study (e.g., sabotage). Therefore, we instead focused on the relevance of individual items rather than full scales.

To determine relevance, items from established scales were compared with the Air Force definition of *maltreatment* or *maltraining*, AETC reports, and other Air Force surveys. Relevant content was reviewed from several measures, including counterproductive work behavior (Gruys and Sackett, 2003; Spector et al., 2006), workplace deviance (Bennett and Robinson, 2000; Stewart, et al., 2009), workplace incivility (Cortina et al., 2001), aggression (Buss and Perry, 1992), abusive supervision (Tepper, 2000), and workplace violence (Rogers and Kelloway, 1997). The majority of the relevant content from these measures targeted violence and aggression, both verbal and physical. To ensure sufficient representation of other forms of maltraining and maltreatment not measured by these constructs, we also reviewed items from surveys previously conducted by the Air Force (e.g., the 2012 U.S. Air Force Academy climate survey).

This process resulted in a preliminary set of 56 items, which were then presented to 23 SMEs at AETC for review. The SMEs included AETC headquarters and BMT leaders as well as other stakeholders (e.g., SARCs, the Office of Special Investigation, chaplains, and medical professionals). We structured this review by asking the SMEs to consider the relative seriousness of each behavioral item and by encouraging them to suggest improvements to each item (e.g., changes to confusing or ambiguous wording, incorrect terms). Based on this feedback, we eliminated 32 items for reasons that included lack of clarity, ambiguous meaning, and redundancy with other items. Other items were then consolidated, resulting in a final set of 18 draft items representative of MTI maltreatment and maltraining, including inappropriate training and discipline, abuse of power, verbal threats and abuse, and physical threats and abuse. We then further refined and reduced these items in steps 3 and 4 of our survey development process, resulting in a total of 17 items for inclusion in the final survey.

Section III: Unprofessional Relationships Between MTIs and Trainees

Although some research literature (e.g., on workplace romance) may appear relevant to this topic, those sources do not generalize well to the types of relationships and interactions prohibited by the Air Force. Therefore, we developed a set of 23 preliminary items based on a thorough review of Air Force policies and reports, previous BMT surveys, and MTI training materials to assess the extent to which MTIs engage in unprofessional relationships with trainees. Items were then refined, eliminated, or consolidated after an internal review to minimize redundancy with other planned scales (e.g., sexual harassment) and to ensure consistency with AETC policies. This process resulted in 16 items, which were then presented to AETC SMEs to gauge the relative seriousness of each item and to provide suggestions for improvement. Based on this feedback, we eliminated several items and refined others, for a total of 13 draft items about unprofessional MTI relationships. We then further refined and added an additional three items based on the feedback in step 3 (additional stakeholder reviews and RAND's quality assurance review) of our survey development process. These 16 items were then refined again based on feedback during our test of the survey items.

Section IV: Sexual Harassment of Trainees by Anyone at BMT

Measuring sexual harassment presents a number of challenges. Definitions vary across research teams and across time as new legal opinions are issued. In the absence of clear methodological guidelines, research teams rely on a variety of validated and novel, as well as nonvalidated, instruments to measure sexual harassment. However, the most commonly used and currently the best validated instrument is the SEQ (Fitzgerald, Gelfand, and Drasgow, 1995). Less commonly utilized measures include the Inventory of Sexual Harassment (ISH) (Gruber, 1992) and the Sexual Harassment Inventory (SHI) (Murdoch and McGovern, 1998). On review, the ISH and SHI were determined not be appropriate for this survey instrument given the substantial overlap with the more widely used and better validated SEQ.

The SEQ is a 28-item, self-reported inventory assessing a range of sexually harassing behaviors. Participants indicate the frequency with which they have experienced, for example, a coworker's "crude sexual remarks," "sexist comments," or "repeated requests for drinks or dinner, despite rejection" (Fitzgerald, Gelfand, and Drasgow, 1995). The measure includes three subscales, each conforming to the three broad categories of sexual harassment (quid pro quo, gender discrimination, and unwanted sexual attention). In civilian samples, the psychometric properties of the scale are strong.

To assess the unique circumstances surrounding sexual harassment of women in the military, the SEQ was revised in 1999 (Fitzgerald et al., 1999). The resulting SEQ-DoD incorporates minor revisions to the language of the civilian SEQ, removes a number of redundant items, and adds new military-specific items suggested by focus groups. The revised instrument also contains four rather than three subscales. The gender discrimination factor was split into two distinct factors: a *sexist hostility* factor, or behaviors that demean a person because of his or her gender, and a *sexual hostility* factor, or behaviors in which sexual content is used purposely to offend a coworker. The remaining factors were the unwanted sexual attention and quid pro quo factors found in civilian samples.

Note that scale scores must be interpreted as "experiences consistent with sexual harassment" rather than "sexual harassment" per se. Sexual harassment is a complex legal construct and many scale items fail to include all indicators necessary to meet the legal standard (e.g., the victim must be offended, and the offensive behavior must meet a "reasonable person" standard, or, failing that, the perpetrator must be aware that the victim is offended and *continue* his or her behavior after learning that the behavior is offensive). Although the scale is the most widely used in the field and considered the best available, it has limited utility for measuring the incidence or prevalence of sexual harassment as it is legally defined. It also may be poorly suited to measuring the sexual harassment of men. Because the SEQ was designed to assess sexual harassment of women, some domains that men typically find offensive are underrepresented (e.g., challenges to one's masculinity and threats to heterosexuality).

A significantly revised 18-item version of the SEQ is currently included in the DMDC's biennial gender relations survey (Rock et al., 2011), and Stark and colleagues (2002) have developed a shortened 16-item version, distinguished with a suffix: the SEQ-DoD-s.

We selected the SEQ-DoD-s as the measure of sexual harassment due to the scale's strong psychometric properties, widespread use, military-specific focus, and reduced length (Stark et al., 2002). A small number of items did not precisely match the BMT environment, however. To address these minor problems, we modified the scale to ensure that all items were appropriate within the BMT context. The changes were as follows:

- Consistent with current Air Force standards, the word *gender* replaced *sex* where appropriate.
- Added two items to better assess behaviors that men, on average, are more likely to find sexually harassing (see Stockdale, Visio, and Batra, 1999):
    - "Called you gay as an insult (for example, 'fag,' 'queer,' or 'dyke')?"

- "Insulted you by saying you were not acting like a real man or real woman (for example, called you a 'sissy' or said you were 'acting like a girl' or 'pretending to be a man')?"

- Replaced outmoded language (*stroke*, *fondle*) with current use (*touch*), as requested by the quality assurance review of the test instrument prior to fielding.
- Revised language in two items to be consistent with the BMT environment (i.e., used "implied you would receive better performance evaluations" instead of "implied faster promotions," and replaced "upcoming review" with "upcoming test").
- Dropped an item that assessed continual pressure for dates, because dating is prohibited during BMT.
- Added the phrase "with him or her" to the item "treated you badly for refusing to have sex."
- Dropped the word *romantic* from the item assessing "unwanted attempts to establish a romantic sexual relationship" to ensure that all sexual relationships were included.

We then further refined and reduced these items in steps 3 and 4 of our survey development process, resulting in a total of 16 final items designed to measure sexual harassment.

Any trainee who answers affirmatively to at least one item indicating an experience consistent with sexual harassment then receives a series of follow-up questions that request details about the most serious event or the event that had the greatest effect on him or her. As described in the main body of the report, although trainees may have experienced more than one event, asking details about each of the events listed would substantially lengthen the survey and could potentially identify victims who would have a much longer completion time. To improve recall, some survey instruments focus respondents on only the most recent event for follow-up questions. For this survey, which prompts recall of only eight weeks of experiences, we did not believe that memory challenges would pose a significant threat. Additionally, we believed that AETC leadership would have a greater need for documentation of the most-serious events. Therefore, we chose to focus on the "most serious" event in the follow-up questions.

These items were modeled on a set of items in the DMDC Workplace and Gender Relations Survey (Rock et al., 2011). Given the unique features of BMT, we modified the original DMDC items and responses for this survey. Follow-up items assess the number, gender, and status (trainee, MTI, other military personnel, or nonmilitary personnel) of the perpetrator(s).

Section V: Trainee Unwanted Sexual Experiences with Anyone at BMT

Measurement strategies that define sexual assault narrowly, surveys that rely on crime reports, and surveys that use the word *rape* tend to produce small prevalence estimates, while those that ask behaviorally specific questions built on the legal definition of sexual assault tend to produce the largest prevalence estimates (Fisher, 2009; Tjaden and Thoennes, 1998). Given little evidence that overreporting is a problem and victims' disinclination to reveal sexual trauma, most investigators have urged reliance on measurement strategies that encourage accurate and full reporting—that is, estimates drawn from participants who were assured confidentiality and who responded to behaviorally specific items are considered to be more accurate (Kruttschnitt, Kalsbeek, and House, 2014).

RAND researchers reviewed all widely used measures of sexual assault, including the Sexual Experiences Survey (Kolivas and Gross, 2007; Koss and Oros, 1982; Koss, Gidycz, and Wisniewski, 1987), the DMDC Unwanted Sexual Contact Assessment (Rock et al., 2011), the National Intimate Partner and Sexual Violence Survey (Black et al., 2011), the National Violence Against Women Survey (Tjaden and Thoennes, 1998), and the National College Women Sexual Victimization Survey (Fisher, Cullen, and Turner, 2000). Because evidence suggests that two-phase assessments beginning with a single-item screen may underestimate the number of sexual assaults (Koss, 1993), we rejected the Unwanted Sexual Contact Assessment for the AETC survey. At 22 items, the National Intimate Partner and Sexual Violence Survey was deemed too lengthy for a survey tasked with assessing a number of other domains, and the National Violence Against Women Survey omitted a number of unwanted sexual experiences that we deemed important to assess in the AETC environment. We considered the Sexual Experiences Survey and National College Women Sexual Victimization Survey but ultimately rejected these scales because the items' complexity required participants to read at a high level.

Given these decisions, it was necessary to create a new measure of sexual assault. We based this new measure on the measures reviewed above with modifications to ensure that the final scale length was ten items or fewer, the required reading level was appropriate, categories of sexual assault were described in behaviorally specific terms, and items were consistent with the UCMJ definition of sexual assault. Due to the prevalence of sexual assault among active-duty men and women (1.2 percent and 6.1 percent, respectively, are sexually assaulted each year; Rock, 2013), we designed the scale to apply to both male and female victims. The final survey measure includes eight items that assess exposure of private areas of the body (requested by AETC), unwanted sexual contact, frotteurism, and attempted and completed oral, vaginal, and anal rape.

All items ask about incidents that occurred *during* BMT. Although it is likely that many trainees enter basic training with a history of sexual victimization (McWhortner et al., 2009), this survey is designed to assess only those experiences that occur while the trainee is under the command of AETC. Items are described in behaviorally specific terms rather than with technical language. For example, rather than asking directly about frotteurism, the trainee item reads: "During BMT, did anyone touch, kiss, or rub up against the private areas of your body when you didn't want them to?" and includes a note defining *private areas* in explicit language to minimize ambiguity and possible variation in interpretation. All items are designed to be consistent with UCMJ definitions of sexual assault.

With the exception of two items assessing vaginal assaults (which male trainees are not asked), all items are equally applicable to male and female victims. Male victims are less likely than females to report assault via official channels (DoD, 2013), so ensuring that the survey assessed victimization among men and women was vital to accurately tracking all sexual assaults. Because of concerns about underreporting and the potential for differing interpretations of the term *consent*, we avoided the language *nonconsensual* or *without your consent* in favor of *unwanted sexual experience*. This language should help capture incidents in which the trainee indicated verbal or physical nonconsent as well as assaults in which nonconsent could not be communicated (e.g., the perpetrator drugs the victim or capitalizes on alcohol intoxication).

Any trainee who answers affirmatively to at least one scale item receives a series of follow-up questions that request details about the most serious event or the event that had the greatest effect on him or her, including the number of unwanted sexual experiences. We chose to focus on the most serious event for the same reasons described in the section on sexual harassment follow-up questions. We also expect that the proportion of trainees who experience multiple unwanted sexual experiences will be small; therefore, for most victims, follow-up questions will be easily mapped onto the single assault they experienced. These items were modeled after a similar set of items included in the DMDC Workplace and Gender Relations Survey (Rock et al., 2011). Given the unique features of the BMT environment, we modified the original DMDC items and responses for this survey. We then further refined these items in steps 3 and 4 of our survey development process.

Trainee Reporting or Telling Others About Abuse and Misconduct

We developed a reporting section for the survey to assess the decisions and experiences of any trainee who either reported or chose not to report an experience with bullying, maltreatment or maltraining, unprofessional relationships, sexual harassment, or unwanted sexual experiences. Trainees who indicated awareness of another trainee who had experienced one of these events were also surveyed about their reporting decisions and experiences.

This section first asks if the trainee reported the incident and to whom. To assess the extent to which bystanders may be aware of events without coming forward, one item asks if the trainee told other trainees about the incident. The next item asks if the trainee reported any of the behaviors in the section to a number of other sources at BMT, including an MTI or someone else in the chain of command, the chaplain, and the SARC. Since trainees also have the option of making a report through an anonymous critique drop box or dorm hotline, we included these options as well.

For trainees who chose not to report the incident, the survey includes a list of potential barriers to disclosure and asks them to select the reasons they did not report any incidents. This section was modeled after a similar set of items in the DMDC Workplace and Gender Relations Survey (Rock et al., 2011). However, given the unique features of the basic training environment, we modified the original DMDC items and responses substantially based on our own expertise in the BMT environment and on feedback from AETC headquarters and BMT leadership, MTIs, and trainees.

Trainees who say that they did report an incident are asked a series of questions regarding their experiences after making a report. This section was also based on a similar set of items in the DMDC Workplace and Gender Relations Survey but was adapted to the BMT environment. We then further refined these items in steps 3 and 4 of our survey development process, as well as based on considerations for packaging and reporting the results.

Trainee Perceptions of Squadron Climate

We reviewed a number of established measures in the scientific literature that are designed to assess areas of organizational climate related to misconduct, including the violence prevention climate (Kessler et al., 2008), safety climate (Zohar, 2014), and ethical climate (Victor and Cullen, 1988; Trevino, Butterfield, and McCabe, 1998). The most-relevant climate measures for this survey focus on sexual harassment and individuals' perceptions of whether the organization tolerates sexual harassment and implements related policies and procedures designed to prevent it (see, e.g., Culbertson and Rodgers, 1997; Williams, Fitzgerald, and Drasgow, 1999). However, the most common measure used to assess what is called "organizational tolerance for sexual harassment" is fairly lengthy and focuses on an office context, making it difficult to adapt to BMT. Furthermore, our goal was to use similar items to assess the climate for each of the abuse and misconduct domains (Hulin, Fitzgerald, and Drasgow, 1996).

Therefore, using the definition for organizational climate in the literature along with definitions of relevant policies for each abuse and misconduct domain as a foundation, we developed items to measure perceptions of the squadron climate for each abuse and misconduct domain. In developing the items, we also reviewed and drew on item construction from other relevant climate measures in the literature (e.g., Culbertson and Rodgers, 1997; Kessler et al., 2008; Trevino, Butterfield, and McCabe, 1998; Williams, Fitzgerald, and Drasgow, 1999). This resulted in nine items to assess perceptions about squadron climate and the extent to which leaders enforce policies and encourage the reporting of incidents. We then further refined and reduced these items in steps 3 and 4 of our survey development process, resulting in a total of four items that repeat for each abuse and misconduct domain in the final survey. It is important to note that the scales are not intended to assess whether shared perceptions of climate exist (i.e., organizational- or group-level climate). Instead, the survey focuses on measuring individual-level perceptions of the climate, also known as the "psychological climate" (Ostroff, Kinicki, and Tamkins, 2003).

Section VI: Trainee Perceptions of BMT Feedback and Support Systems

One recommendation after sexual misconduct was detected at AETC was to increase the visibility of supervisors, commanders, chaplains, and SARCs in the training environment. It was expected that increasing trainees' familiarity with these individuals would facilitate the reporting of sexual abuse and other misconduct. We designed this section of the survey to serve two purposes: First, trainees are asked how easy it would be to contact 12 different people in positions of responsibility at BMT (e.g., first sergeant, MTI, chaplain, law enforcement) to make a report about abuse and misconduct. Second, to assess the success of efforts to increase the visibility of individuals to whom a report could be made, trainees are asked whether they would recognize each of those same individuals (excluding the BMT support personnel they may rarely come into contact with; ten items total). We identified the individuals on this list through discussions with AETC headquarters and BMT leadership and support staff.

In addition, we developed 12 separate items to assess trainees' perceptions of the available reporting or feedback systems at BMT. Again, these items were developed based on our discussions with AETC headquarters and BMT leadership and support staff regarding the different reporting options that trainees had available to them. We then further refined these items in steps 3 and 4.

MTI Survey

Sections I–V: MTI Quality-of-Life Measures

The MTI survey contains several sections designed to measure MTI quality of life. Prior to the current project, AETC already had the QOLS, which was administered to MTIs once a year. To prevent survey burnout among MTIs, AETC asked RAND to review the QOLS and work to integrate the constructs measured in the QOLS into RAND's survey so that only a single survey would be given to MTIs. A review indicated that the QOLS items measured a wide range of constructs, including job attitudes, work stress, and perceived leadership effectiveness. The QOLS also asked for background information (e.g., marital status) that was not considered for the RAND survey, since these items would decrease the desired level of respondents' anonymity. In addition to the constructs identified in the QOLS, we also sought out established scales and developed new items to address other topics raised in AETC reports and in meetings with AETC headquarters and BMT leadership and MTIs.

Section I: Job Attitudes

As described in the main body of the report, job attitudes have been shown to be useful indicators of employees' well-being and their job performance (Cooper-Hakim and Viswesvaran, 2005; Meyer et al., 2002; Meyer and Maltin, 2010; Riketta, 2002). Not only were items measuring commitment already included on the QOLS but we concluded that organizational attitudes would also be consistent with AETC's broader goals to create a positive and safe work environment for MTIs and trainees. We chose to focus on two of the most common general job attitudes for inclusion in the survey: *organizational commitment* and *job satisfaction*.

Although there are different conceptualizations of organizational commitment (e.g., Jaros et al., 1993; Mayer and Schoorman, 1992; O'Reilly and Chatman, 1986), most overlap to some extent with the three-component model of organizational commitment developed by Allen and Meyer (1990). This model distinguishes among three well-known facets of commitment: affective, normative, and continuance commitment. Extensive research conducted on these facets shows that affective commitment demonstrates the strongest and most-consistent links with various organizational outcomes (e.g., attendance, performance, stress; Meyer et al., 2002), while normative and continuance commitment generally demonstrate weaker relationships with these criteria; therefore, we considered only items measuring affective commitment for inclusion in the MTI survey. The scales developed by Allen and Meyer (1990) and Meyer, Allen, and Smith (1993) to measure organizational commitment are well established, having been used in more than 100 studies (see Meyer et al., 2002). Consequently, we adapted items from these scales as the foundation for measuring affective commitment on the MTI survey. Several of the items from this scale are also similar to items that were previously included in the QOLS.

There are two broad approaches to the measurement of job satisfaction (Judge et al., 2001). One approach is to focus on global job satisfaction (e.g., asking, "Overall, how satisfied are you with your job?"). Another approach is to focus on various facets of job satisfaction (e.g., satisfaction with supervisor, coworkers, work itself). Both approaches offer different advantages. On the one hand, global measures of job satisfaction not only are simple but can also serve as stronger indicators of relevant criteria, such as life satisfaction and happiness (Bowling, Eschleman, and Wang, 2010). On the other hand, facet measures of job satisfaction are more diagnostic of potential problems within an organization. That is, global measures of job satisfaction may indicate that employees are generally unhappy, but these measures do not provide organizational leaders with the diagnostic information necessary to target specific areas for improvement (e.g., coworker conflict). Rather than restricting the MTI survey to one approach, and because both approaches have value, we drafted items to represent global job satisfaction as well as the facets most relevant to MTIs. However, we did not include facets that would generally be beyond the control of BMT leaders (e.g., satisfaction with pay).

To assess global job satisfaction, we searched the research literature for a well-established and psychometrically sound measure and identified a one-item measure developed by Highhouse and Becker (1993) for inclusion in the survey. The survey then assesses facet measures of job satisfaction most relevant to MTIs with the scales described in the section on the work environment.

Section II: Work Environment

Several facet measures of job satisfaction have been developed; however, many of these are proprietary instruments sold for profit or contain too many items for practical use on the MTI survey. For example, the Job Descriptive Index not only is a proprietary instrument but it also contains 72 items to measure five facets of job satisfaction, including work on present job, present pay, opportunities for promotion, supervision, and coworkers.[10] Therefore, we decided to focus on developing measures that were consistent with these widely accepted facets, but that were also most relevant to the MTI context. Based on the reports and feedback we received in step 1 of the survey development process, our knowledge of relevant job satisfaction facets, and

---

[10] See the "Job Description Index" page on Bowling Green State University's website: http://www.bgsu.edu/arts-and-sciences/psychology/services/job-descriptive-index.html

topic areas included as part of the QOLS, we identified the following topic areas to assess: *organizational support, MTI interpersonal treatment, MTI perceptions of trainees,* and *assignment and promotion opportunities.*

Several different established scales designed to measure organizational support, or often the absence of organizational support in the form of work overload, exist in the research literature. However, we found that these measures did not necessarily capture the key issues important for the MTI context. Therefore, we developed an initial set of six draft items designed to specifically assess MTI satisfaction with work resources in significant areas raised in our discussions with key stakeholders and reviews of reports. The items also assessed similar topic areas measured in the QOLS. We then further refined and reduced these items in steps 3 and 4 of our survey development process, resulting in a total of four items for inclusion in the final survey.

To measure MTI interpersonal treatment, we identified an established and validated measure in the literature by Donovan and colleagues (1998) that we were able to adapt to the BMT context. Based on our review of AETC reports and discussion with BMT leaders, we decided to make a distinction in this section between seasoned MTIs and MTIs more generally to capture the power differential between more experienced, senior MTIs and less experienced, younger or *rookie* MTIs. Therefore, we adapted four items from the measure developed by Donovan, Drasgow, and Munson (1998) to first assess how well seasoned MTIs treat rookie MTIs at BMT and then repeated those same four items to assess how well MTIs treat each other more generally.[11] These items were then reviewed in step 3 for clarity, and then we confirmed the factor structure and reliability of the scales in step 4.

One area that is somewhat unique to an MTI's satisfaction is the quality and integrity of trainees, which was raised as a concern in AETC reports and MTI reviews of an initial survey draft. Therefore, we developed four initial items to monitor MTI concerns about trainees. We then further refined and added to these items in steps 3 and 4, resulting in a total of six items for inclusion in the final survey.

Finally, consistent with other facet measures of job satisfaction (e.g., Kinicki et al., 2002), we developed two items to measure MTIs' perceptions of their promotion and assignment opportunities during and following their assignment at BMT. We then further refined the wording of these items in steps 3 and 4.

Section III: Leadership

Based on our discussions with AETC leadership, reviews of related BMT abuse and misconduct reports, and items contained in the previous QOLS, we identified two general areas to assess in terms of leadership at BMT: *general leader treatment of MTIs* and *leader ethical conduct.*

---

[11] Donovan, Drasgow, and Munson's scale is proprietary and was used with the authors' permission.

To measure leader treatment of MTIs, we adapted a leader-focused 14-item scale from the same established measure used to assess MTI interpersonal treatment (Donovan, Drasgow, and Munson, 1998).[12] This measure contained items relevant to the context and issues facing MTIs at BMT, included items similar to those asked on the previous QOLS, and had been validated in the empirical research literature. We then further refined and reduced these items in steps 3 and 4 of our survey development process, resulting in a total of ten items for inclusion in the final survey.

To measure MTI perceptions of leader ethical conduct, we adapted items from a well-established measure of ethical leadership (Brown, Trevino, and Harrison, 2005) that has been used in other military contexts, such as the Army (Schaubroeck et al., 2012). Although the original measure includes ten items, we selected only five items for inclusion on the draft survey due space limitations and applicability to the BMT context. We then further refined the item wording and confirmed the factor structure and reliability of the scale in steps 3 and 4.

Section IV: Work and Family Stressors

Based on our discussions with AETC leadership, feedback from MTIs, and reviews of related BMT reports on items on the QOLS, we designed two separate sections to measure stress. The first section addresses the extent to which MTIs' *work and family life* affect each other. The second section attempts to identify other *specific work-related stressors* that affect MTIs. It is important to note that the survey does not attempt to capture every possible source of stress for MTIs. Because the MTI survey is already lengthy, we focused on only those stressors that emerged as the most relevant for MTIs.

To assess the extent to which MTIs experience work-family conflict, we reviewed the research literature for measures with strong psychometric properties that could also be easily adapted to the MTI context. We identified several potential measures (e.g., Gutek, Searle, and Klepa, 1991; Kopelman, Greenhaus, and Connolly, 1983; Small and Riley, 1990; Stephens and Sommer, 1996), but ultimately settled on an established measure by Carlson, Kacmar, and Williams (2000) that had good psychometric properties and was suitable to the MTI context. The full measure by Carlson and colleagues contains six total dimensions of work-family conflict focused on strain, time, and behavior interference between work intruding on family and vice versa. However, we found that many of the items in the time and behavior subscales were not as relevant for the BMT environment or were already assessed elsewhere on the survey (e.g., we decided to include items on spending too much time at work in the section on specific stressors). Therefore, we chose to focus only on the strain-based interference subscales in the survey, which included items similar to those found on the previous QOLS. We included a two-dimensional scale, with three items measuring strain-based family interference with work (e.g., "Due to stress at home, I am often preoccupied with family matters at work") and three items measuring strain-

---

[12] This scale is proprietary and was used with the authors' permission.

based work interference with family (e.g., "When I get home from work, I am often too worn out to participate in family activities or responsibilities"). We confirmed the factor structure and reliability of the scale in steps 4.

To assess the specific stressors MTIs might face, we followed an approach proposed in previous research (e.g., Mahmood et al., 2010; Shane, 2010) and asked MTIs to indicate the extent to which various items have caused stress over the past six months. To identify which stressors to include on the survey, we reviewed AETC reports and previous Air Force surveys and held meetings with AETC headquarters and BMT leaders and MTIs. We identified 26 initial potential work-related stressors for inclusion on the test survey. Although constructed to represent areas of particular concern to BMT, many of these stressors have been highlighted in other research in both military and civilian contexts. For example, several stressors reflect different forms of role stress, such as role ambiguity and role conflict (e.g., conflicting job expectations). Other stressors, such as trainee comment forms, were unique to BMT. The test of the survey also included an "other" response option in which participants could provide a free-text response of additional stressors that were not listed. Based on feedback we received in steps 3 and 4 of the survey development process, we ended up dropping several items and adding others, resulting in a total of 28 stressors on the final survey.

Two follow-up questions also ask MTIs about their work and sleep habits. Although the section on specific stressors includes items that assess MTI stress related to being overworked, we also sought to obtain more-quantitative measures of the amount of time MTIs work and are able to sleep. The survey includes a single question on the average number of hours worked in a day, a single item on the number of hours of sleep an MTI is able to obtain in a 24-hour period, and a single item on overall sleep quality.

Section V: MTI Professional Development

Finally, we also included several items assessing MTI professional development that came directly from the QOLS and were important to continue to track for BMT's training purposes. One question asks MTIs how much their instructional skills have improved over the past six months; another asks if the individual has taken a deliberate development course since becoming an MTI and the extent to which the course was beneficial.

Sections VI–XI: Abuse and Misconduct Toward Trainees

To complement the trainee survey, the MTI survey includes scales to assess MTI *awareness* of bullying, maltreatment and maltraining, unprofessional relationships, sexual harassment, and unwanted sexual experiences in the BMT environment. The content of these scales mirrors the content of the trainee survey. As described in the main body of the report, RAND researchers considered assessing MTIs' self-reported engagement in misconduct, but feedback from MTIs and AETC leadership suggested it was unlikely that MTIs would feel sufficiently protected to reveal personal misconduct. So instead of assessing MTIs' personal misconduct, we adapted items on the trainee survey to assess whether MTIs were personally aware of specific misconduct at BMT in the past six months. Except for those minor adaptations, the scales assessing bullying, maltreatment and maltraining, unprofessional relationships, and sexual harassment are the same for the trainee and MTI surveys.

In contrast, the scale assessing unwanted sexual experiences among trainees was trimmed down for the MTI survey, because MTIs were unlikely to have enough information to answer certain questions. That is because MTIs are told not to press trainees who report a sexual assault for a detailed description of the assault but to focus instead on obtaining support and services for the trainee. The MTI scale of unwanted sexual experiences was reduced to four items to assess whether MTIs were personally aware of the following unwanted sexual experiences among trainees: exposure of private areas of the body; unwanted sexual contact or frotteurism; oral, vaginal, or anal sexual assault; and attempted oral, vaginal, or anal assault.

MTI Perceptions of Squadron Climate

The MTI survey also includes scales to measure perceptions of the squadron climate related to each of the abuse and misconduct domains: bullying, maltreatment and maltraining, unprofessional relationships, sexual harassment, and unwanted sexual experiences. The content of these scales mirrors the content of the trainee survey.

MTI Reporting Norms

We also specifically developed items to measure perceptions of the general reporting environment for MTIs or norms for reporting behaviors. RAND researchers considered assessing MTIs' self-reported actions of reporting or not reporting incidents of which they were aware and potential barriers for why they chose not to report an incident. However, feedback from MTIs and AETC leadership convinced us that it was unlikely MTIs would feel sufficiently protected to reveal that they failed to report an incident. Instead, we determined that the best approach would be to ask MTIs to report whether they perceived MTIs in general as being willing to report an incident or norms for MTI reporting behaviors. Based on discussions with AETC headquarters and BMT leadership as well as MTIs, we developed an initial set of nine items to assess MTI reporting norms. We then further refined and reduced these items in steps 3 and 4 of our survey development process.

Clarity of Abuse and Misconduct Policies

Finally, we developed a series of items to assess the clarity of policies related to each of the abuse and misconduct domains on the survey. Like the other sections, these items were further refined and reduced in steps 3 and 4, resulting in the final five items on the final survey.

Step 3: Review and Refinement of Draft Surveys

After developing initial drafts of the trainee and MTI surveys, we held review and feedback sessions with three key groups to refine the draft surveys: (1) AETC leaders and other key stakeholders and SMEs (e.g., a SARC, a chaplain, Air Force law enforcement), (2) MTIs, and (3) trainees. The AETC leader and stakeholder group included 14 individuals. The trainee group included approximately 25 trainees total, with roughly equal representation of male and female trainees in their seventh or eighth week of BMT, and the MTI group included approximately 30 MTIs total, with roughly equal representation of male and female MTIs. To account for potential differences in the language or directions used in different squadrons, we randomly selected trainees and MTIs from across squadrons to participate in the meetings. Although all selected participants came to the feedback sessions, participation in the discussion was voluntary, and some trainees and MTIs chose to be less involved.

The goal of these review and feedback sessions was to gather input about the survey items (e.g., content, readability), survey length, order of items, and practical issues regarding survey administration. For each session, a member of the project team explained the purpose of the survey and the goal of the meeting. He or she then passed out draft copies of the survey to participants, who were instructed to read through it section by section and provide written comments. The AETC leader and stakeholder group reviewed both the trainee and MTI surveys, while trainees and MTIs reviewed only the survey relevant to them. After the participants reviewed each section independently and provided written comments, the project team members led a discussion of the section as a group. We held separate meetings for the female and male trainees and female and male MTIs given the sensitivity of the survey topics.

Feedback from the sessions included comments in the following areas:

- proper BMT language or terminology
- items that were confusing
- items or questions that may not be applicable to BMT
- missing topics from the survey
- willingness to be open and honest in answering the survey questions.

These client-based review sessions provided valuable feedback on the survey content and helped further guide revisions to the individual survey items to ensure that the items were applicable to BMT and are understandable. The reviews also helped identify missing content; we added a number of measures to the MTI survey to better assess MTI quality of life. Given the significant changes made to the MTI survey after these sessions, we also held one follow-up session with a group of MTIs to review the revised survey.

Following these review sessions, we then held a RAND internal quality assurance review session for the trainee survey with three independent reviewers: one internal RAND statistician with expertise in survey development and analysis, one internal RAND researcher with expertise in military sexual assault and survey research, and a third external reviewer who holds a law degree, has been an attorney for 28 years, and served as the chief of the Air Force Sexual Assault Prevention and Response Office from 2005 to 2007.

The quality assurance review consisted of a daylong interactive session with all three reviewers in which the project team presented an overview of the study goals, the draft trainee survey, and the proposed administration methods. The reviewers were provided a copy of the survey prior to the session and were given an opportunity to ask clarifying questions of the study team. The reviewers then provided jointly written feedback to the project team that outlined key concerns and recommendations for improving the trainee survey and administration procedures. Recommendations focused on:

- minimizing the survey length
- timing the administration to avoid survey fatigue
- refining the survey instructions to promote open and honest responses
- refining the response scales
- clarifying language used in some survey items
- adding follow-up questions for individuals who experienced an incident.

Based on this feedback, the project team revised the trainee survey again and carried over relevant revisions to the MTI survey.

Step 4: Test of the Surveys

Although considerable effort was invested in selecting survey measures and designing new BMT-specific items, it is critical to test a survey with the intended population before disseminating it for widespread use. Testing serves a number of instrument-construction purposes. It lets investigators test the performance of each individual item and scale and potentially eliminate items that are endorsed too rarely or are duplicative of other items. Shortening the survey also reduces the time and costs necessary to field a survey—particularly for a survey that will be given repeatedly. Examining the comments of test participants can also help identify items that require clarification or may not be relevant to the specific environment. Moreover, comments may identify relevant topics or attitudes that were overlooked and should be added to the survey. Finally, testing the survey allows investigators to confirm the psychometric properties of existing scales when they are applied in a new setting with a new population.

Below, we provide details on the test RAND conducted to help refine the BMT trainee and MTI surveys. We had three goals: (1) revise the content of the final surveys, (2) examine and collect data on when and how to best administer the surveys, and (3) provide initial baseline data for AETC. The test took place over a three-day period in July 2013.

Participants

Trainees

To examine the potential impact of administering the survey at different points in trainees' progression through BMT, we administered the survey to trainees from different week groups. We gave the survey to all trainees who had completed weeks three or seven of BMT at the time the survey was administered, as well as a random sample of students who had just graduated from BMT and were starting TT at Lackland Air Force Base.

A total of 1,262 trainees attended the survey sessions. Of these, 1,138 (90 percent) consented to completing the survey as part of testing procedures. However, a small proportion of those who completed the survey were dropped from the sample due to evidence that their responses may not be valid. Ninety-three (7 percent) were dropped from the final sample because their responses were incorrect on several screening questions designed to detect random or inattentive responding (e.g., the trainee responded "Never" to an item that read "Please select 'Daily' for this item"). Three participants were dropped because they indicated that they were "not at all open and honest" on at least some responses.

Of the 1,042 trainees included in the final sample for analysis, 787 (76 percent) were men, 251 (24 percent) were women, and four did not indicate gender. In terms of week group, 463 trainees (44 percent) had just completed their third week of training, 490 (47 percent) had completed their seventh week, and 89 (9 percent) were in their first week of TT.

MTIs

All MTIs stationed at Lackland Air Force Base who had been in the role for at least one month were invited to participate in the survey test. Of the 416 MTIs at Lackland during the dates of the test, 308 (74 percent) attended a survey session. Of those who attended a session, 280 (91 percent) agreed to participate in the study. Forty (14 percent) of those who completed the survey were dropped from the final sample due to evidence of invalid responses. Thirty-nine were screened out because they gave incorrect responses on several screening questions designed to detect random or inattentive responding. One MTI was dropped for indicating that he or she had been "not at all open and honest." The final sample of 240 MTIs represents 58 percent of all MTIs at Lackland at the time of the survey administration.

Of the 240 MTIs in the final analytical sample, 17 (7 percent) had been at BMT for six months or less, 66 (28 percent) had been there for seven months to two years, 155 (65 percent) for two or more years, and two did not reveal their time at BMT. The majority of participants were line MTIs (n = 160, 67 percent), and the remaining participants were either supervisors (n = 31, 13 percent) or in other roles (n = 46, 19 percent) (e.g., Warrior Skills and Military Studies [WS/MS], Basic Expeditionary Airman Skills Training [BEAST] cadre). Three chose to not identify their primary duties. To protect the confidentiality of the small number of female MTIs, gender was not assessed.

Procedures

Trainees

The trainee survey was administered in one of two BMT classrooms. Each classroom was outfitted with 120 computer workstations. Because of the classroom size, two flights were scheduled for 60-minute sessions. RAND was responsible for and conducted the administration. At the start of the session, a RAND researcher read the informed consent, provided an opportunity for questions, and then instructed trainees to open the survey on their workstations and begin.

When trainees opened the survey window, the first screen was the informed consent statement that the researcher had read aloud. Trainees were provided with three options: (1) consent to participate, which launched the survey instrument; (2) decline to participate, which closed the survey instrument; and (3) decline to participate but click through the survey without answering the items so other trainees would not know they had declined to participate. Although we expected most trainees to complete the survey before 60 minutes had elapsed, all trainees remained in the classroom until the end of the session since they are not allowed to move around BMT freely without an MTI. RAND researchers instructed them to wait quietly or read their study materials until the end of the survey session.

During the survey administration sessions, the research team implemented a variety of procedures to assure trainees that their responses were truly confidential. For training purposes, BMT does not afford trainees a great deal of privacy, and trainees come to expect that their behavior and performance are monitored. However, for this survey, it was vital that trainees felt confident that their responses were protected and confidential. To instill this confidence, we took a number of steps. First, no MTI or Air Force official could enter the survey room during the administration period. Second, with the support of Lackland computer specialists, all workstations were programmed for use without the normal CAC, so trainees did not have to log on to the stations in a way that would identify them. Third, demographic questions were limited to the bare minimum to limit the risk of identification by inference. Finally, all workstation monitors were outfitted with privacy protectors, which limit the ability of anyone not seated directly in front of the computer to view the screen's content.

MTIs

MTI surveys were administered in the same BMT classroom used for trainees. As in the trainee session, a RAND researcher read the informed consent aloud, provided an opportunity for questions, and then instructed MTIs to open the survey instrument. The first survey screen included the informed consent and provided the option to consent and proceed through the survey or to decline to participate. MTIs who declined to participate were allowed to leave the survey session. The same procedures used to protect trainee confidentiality were implemented for MTIs as well (no Air Force officials in the classroom, no CAC card log-in, limited demographic items, and computer privacy protectors).

Trainee and MTI Scale Analysis and Item Reduction

For the test versions of the surveys, we included more items than we intended to include in the final versions of the survey since many of the scales were developed specifically for the BMT context (see Tables A.1, A.2, and A.3 for notes on the number of items dropped). This allowed us to assess the performance of individual items and scales and retain only those items that helped create reliable and parsimonious measures of the intended underlying constructs on the survey. We eliminated items that were endorsed too rarely, were not worded clearly enough, were duplicative of other items, or did not measure the intended underlying construct. In addition, testing the survey allowed us to confirm the psychometric properties of existing scales in the BMT context.

The final analytic data sets had largely complete-case data (i.e., participants responded to all presented questions) with a relatively small degree of missing data. Among the questions composing each core section of the trainee survey (i.e., bullying, maltreatment and maltraining, unprofessional relationships, sexual harassment, and unwanted sexual experiences), between 86 percent and 93 percent of the trainee sample completed all presented items within a given section. For the MTI survey, the completion rates were similar, with 84 percent to 93 percent of the MTI sample completing all presented items within a given section.

Analytical Approach

Using these analytic data sets, we next conducted what is known as *factor analysis* for each proposed scale. The goal of these analyses was to identify the collections of items (i.e., scales or domains) that are best represented by single factors and reduce the number of items when possible to help minimize survey length. For this aim, factor analysis is a widely used technique for evaluating the utility of survey items and is useful for developing new survey measures. Simply put, factor analysis is a means of evaluating the extent to which collections of items on a survey measure the same or different underlying constructs.[13] Factor analytic models may be broadly classified as either *confirmatory* or *exploratory* (CFA and EFA, respectively). CFA is commonly used when evaluating previously established scales. A single-factor CFA model tests the hypothesis that the relationships among item responses within a scale can be explained by a single underlying construct. When the single-factor model closely fits the data, the scale is referred to as being *unidimensional*, and it is appropriate to generate a single score from the set of items (e.g., by summing or averaging the response to the set of items). In some situations, the fit of a unidimensional model can be improved by modeling additional factors for small subsets

---

[13] The *factor* or *dimension* is the content being measured by the set of items.

of items. *Local dependence* is when an excess relationship is exhibited by only two items, and a residual correlation is often modeled. In most practical applications, dropping a single item from the item pair will result in the desired unidimensionality of the scale.

In situations where items are newly developed and have not been previously tested, or when a set of items may be represented by multiple factors, it is often useful to begin the analytic process using EFA models. EFA approaches to item analysis are similar to one-factor CFA models but allow more than one factor to emerge from the data. In these situations, additional factors may be caused by subsets of items having shared characteristics (e.g., overly similar content, shared or similar phrasing, or valence [i.e., positively and negatively worded items]). When an EFA suggests that multiple factors are present, it is inappropriate to generate a single score from the collection of items. Rather, each factor should be scored separately or items should be removed from the scale such that the remaining items achieve unidimensionality.

For all proposed scales in the surveys, we conducted the appropriate factor analyses (i.e., CFA for established scales and EFA for newly developed scales) in Mplus computer software (Muthén and Muthén, 1998–2010) with weighted least square means and variance adjusted estimation appropriate for categorical response data.[14] We evaluated the appropriateness of the models based on the following factors: (1) model fit indices (root mean square error of approximation [RMSEA] ≤ 0.08, Tucker Lewis index [TLI] ≥ 0.95, comparative fit index [CFI] ≥ 0.95; Hu and Bentler, 1999; Browne and Cudeck, 1993); (2) the magnitude of the factor loadings; and (3) modification indexes. When these indicators suggested poor model fit (in other words, when the model was not unidimensional), we established revised models that accounted for local dependence (i.e., the phenomena of a pair or cluster of items being more related than would be expected given the underlying dimension). Often, these revised models accounted for local dependence by including residual correlations for pairs of offending items or by establishing additional factors from which composites may be created. During the process of identifying these models, we also removed items that provided evidence of nuisance dimensionality or that provided little additional gain in scale reliability (e.g., items with low factor loadings). In the results that follow, we provide the model fit indexes for all final CFA models; however, because of the many factor analytic models that were evaluated, it is impractical to report the factor loadings for each measure.[15]

---

[14] EFAs were conducted with Crawford-Ferguson rotation and oblique factor extraction (Crawford and Ferguson, 1970).

[15] As part of our assessment, we also reviewed additional statistics related to these items and scales (such as percentages, means, and ranges); however, we do not present them here for human subjects protection purposes.

Trainee Survey Results and Final Item Selection

Abuse and Misconduct Scales

We fit unidimensional factor analytic models to the bullying, maltreatment and maltraining, and sexual harassment scales. The unprofessional relationships and unwanted sexual experiences scales had items that lacked the variation required to permit factor analyses. Results from these models suggested that each scale exhibited local dependence among a small subset of items. For example, modification indexes from an initial one-factor model fit to the sexual harassment measure indicated local dependence between two items related to sexist remarks and put-downs. A subsequent model estimated residual correlation between the locally dependent items. In these situations the item with the weaker loading was generally removed. After accounting for this local dependence in subsequent models by removing small subsets of items from each scale, results showed strong and reliable unidimensional scales (see Table A.1).

Table A.1. Abuse and Misconduct Scale Item Selection

Scale	Items Removed	Items Retained	Reliability	RMSEA	CFI, TLI
Bullying	3	6	0.75	0.045	0.994, 0.991
Maltreatment and maltraining	1	17	0.76	0.019	0.984, 0.982
Unprofessional relationships	0	16	N/A	N/A	N/A
Sexual harassment	1	16	0.74	0.036	0.960, 0.949
Unwanted sexual experiences	0	17	N/A	N/A	N/A

NOTE: Reliability was calculated using Cronbach's alpha, a measure of the internal consistency of the responses to the scale's items. Due to a lack of response variation, reliabilities could not be computed for two of the scales.

Squadron Climate Scales

We developed identical measures of squadron climate for each abuse and misconduct domain. Initial factor analytic models fit to these scales indicated that each scale was represented by two separate factors that corresponded to the positive or negative valence of the items. That is, in each scale, one factor contained the positively worded items and the other factor contained negatively worded items. Because a single factor could not account for the relationships among the items, it would be inappropriate to generate a single score from the entire set of items in each scale. In these situations, a straightforward method for obtaining a unidimensional set of items is to remove the items from one of the two factors. Because the content of the positively worded items was more substantively relevant, we next removed the negatively worded items and fit a unidimensional model to each remaining subset of items. This one-factor model suggested that the positively worded items were strongly unidimensional. However, one additional item that consistently underperformed in each scale was removed, resulting in a four-item climate scale for each core abuse and misconduct domain (see Table A.2). After selecting four-item subsets from

80

each climate scale, we also conducted a five-factor EFA to evaluate the utility of separate climate scales. A five-factor solution resulted in factors defined by the respective climate scales, indicating that each climate domain accounts for unique variance and should be assessed separately. A final five-factor CFA based on the EFA factor structure indicated excellent model fit (RMSEA = 0.057, CFI = 0.994, TLI = 0.992).

Table A.2. Squadron Climate Scale Item Selection and Reliability

Scale	Items Removed	Items Retained	Reliability
Bullying	5	4	0.90
Maltreatment and maltraining	5	4	0.92
Unprofessional behavior	5	4	0.91
Sexual harassment	5	4	0.92
Sexual assault	5	4	0.92

NOTE: Reliability was calculated using Cronbach's alpha.

Remaining Individual Items

The remaining survey items were not part of a scale but rather simple stand-alone items (e.g., for participants who did not report an incident, "please select the items below that describe why you did not report any incidents"). We generated descriptive statistics for each of the remaining items and examined frequencies to identify item responses that were candidates for deletion. For example, if less than 1 percent of trainees endorsed a response option, we considered the appropriateness of dropping the response option from the final survey instrument. We also examined free-text responses for items that included an "other" response option. If survey participants consistently added a new response that had not been included in the original items, we considered the utility of adding the new response option to correct the omission. Based on the results, we ended up dropping several items and adding others.

MTI Survey Results and Final Item Selection

Quality-of-Life Scales

The MTI quality-of-life section of the survey included both established scales and scales that had been developed specifically for this survey. When analyzing the established scales, we first attempted to model the factor structure indicated by how the survey's items are scored. For example, if the established scale were unidimensional, we would initially fit a one-factor model. If model fit indexes and the magnitude of the factor loadings indicated that a single factor was inadequate, we would subsequently fit a more complex model to account for the relationships among the item responses (e.g., EFA, multifactor CFA, or one-factor models with correlated residuals). When we were unable to replicate the factor structure from an established scale, we conducted analyses aimed at obtaining a unidimensional set of items measuring the intended content of the scale. From this set of items, we often selected a subset of items to maintain adequate reliability, while removing items with low factor loadings (i.e., we removed items that did not contribute to the reliability of the scale's scores). Note that for three of the unidimensional measures (treatment of rookie MTIs, general MTI interpersonal Treatment, and ethical leadership), the RMSEA suggests poor model fit, while the incremental fit index, CFI, indicates close model fit. In addition, the scale reliabilities and factor loadings for these models all failed to provide any indication of model misfit. In these cases, it is possible that the χ^2 is overpowered and the RMSEA, which based on the χ^2, is negatively affected, while the CFI remains unaffected. These results often lead researchers to incorrectly assume model misspecification (Miles and Shevlin, 2007). Based on the collection of fit indexes and magnitudes of factor loadings, we suggest that any misspecification in these one-factor models is likely trivial. Table A.3 presents the final item selection, reliability, and model fit indexes for each scale.

Table A.3. MTI Quality-of-Life Scale Item Selection and Reliability

Scale	Items Removed	Items Retained	Reliability	RMSEA	CFI, TLI
Organizational commitment	3	5	0.82	0.039	0.998, 0.997
Treatment of rookie MTIs	0	4	0.74	0.313	0.940, 0.819
General MTI interpersonal treatment	0	4	0.77	0.453	0.919, 0.757
Leader treatment of MTIs	4	10	0.92	0.093	0.993, 0.988
Ethical leadership	0	5	0.87	0.085	0.996, 0.992
Work-family conflict (two-factor model)	0, 0	3, 3	0.89, 0.94	0.105	0.998, 0.995

NOTE: Reliability was calculated using Cronbach's alpha.

Factor analysis of three of the scales (organizational support, perceptions of trainees, and assignment and promotion opportunities) indicated that the items did not measure a single underlying construct and should be analyzed separately.

To help refine and reduce the number of items listed under potential job stressors (which were not intended as a scale), we generated descriptive statistics for the items and examined the interitem correlations to help identify items that were duplicative or had very low endorsement of being stressful. We then considered whether it was appropriate for these items to be removed. The test of the survey also included an "other" response option in which participants could provide a free-text response of additional stressors that were not listed. If survey participants consistently added a new response that had not been included, we considered the utility of adding the new response option to correct the omission. Based on the results, we ended up dropping several items and adding others.

The MTI survey included abuse and misconduct items and squadron climate items that corresponded to the final trainee survey. To maintain consistency, we kept for the final MTI survey the items that aligned with those on the trainee survey. However, we confirmed that the scale reliabilities were still sufficient when using these items with the MTI population (Cronbach's alpha ranged from 0.85 to 0.92).

For each of the abuse and misconduct domains on the MTI survey, we also included items about MTI reporting norms and the clarity of policies. We conducted EFAs on these scales but did not find strong, clearly differentiated content domains across the reporting scales or sufficient reliability for the scales assessing clarity of policies. Because we thought it important to still keep separate items for MTI reporting for each abuse and misconduct domain, we retained four of the original six items on MTI reporting norms to be analyzed separately instead of as a unified scale. We also retained a single item for each abuse and misconduct domain assessing the extent to which MTIs view the related abuse and misconduct policy or law as clear.

Survey Test Feedback Questions

For the purposes of the test, both the trainee and MTI surveys end with a series of items to assess the mechanics of the questionnaire and administration procedures and the participants' willingness to engage in the process.

The first item on both surveys asks participants to indicate how open and honest they felt they could be while answering the survey questions. This item served three purposes. First, it can serve as an indicator of trust in BMT leadership and the organization as a whole. Second, it provided a test of the administration procedures, which were designed to improve confidence in the anonymity of responses and perceived safety in providing honest feedback to leadership. If a high proportion of participants answered this question affirmatively, it would indicate that the administration procedures were successful. If a small proportion answered affirmatively, then additional strategies to improve confidence would need to be implemented. Third, it allowed the researchers to remove the answers of participants who indicated that they were not open or honest in their responses from the set of questionnaires to be analyzed. When participants who do not feel safe providing truthful feedback are included in a survey sample, they can artificially bias downward the estimate of misconduct; therefore, they must be excluded to increase confidence in observed frequencies of events. For this reason, we also kept this item in the final version of both surveys to serve as a screening question for inclusion in analyses.

Both surveys then ask a set of questions regarding how comfortable being open and honest the participant would be under different administration conditions, such as differences in who administered the survey, when it was administered, and whether participants would be identifiable. Given that this survey will be administered by Air Force professionals in the future, RAND included these items to provide feedback about how administration changes might affect future studies. We discuss the results from these questions in more detail in Chapter Four.

Finally, the surveys also included several open-ended questions to allow trainees and MTIs to provide unscripted feedback. One asked about any behaviors related to abuse or misconduct that were not included in the survey that the respondents think should be included. Another asked whether any questions or sections of the survey were confusing or hard to answer. The MTI survey included a third question asking if participants would like to say anything more about their working conditions, leadership, or quality of life as an MTI. The feedback from these open-ended questions was used to further refine some of the wording of the survey items.

Conclusion

Following the test of the surveys and subsequent revisions, AETC headquarters, BMT leadership, and other key stakeholders again provided a final review. Feedback from this review resulted in additional minor revisions to some item wording and directions. The surveys presented in Appendixes B and C reflect the fully revised surveys based on the above development steps.

We do not recommend using the trainee test survey results collected as a part of this development process as a baseline, primarily because our test sample included trainees at different points in the training cycle and because AETC began administering the revised version of the survey only a few months later on a weekly basis.

The initial MTI test survey results, however, can serve as rough baseline for future data collection efforts at AETC. Although some items were revised, dropped, or added, there are significant portions of the survey that remained relatively consistent. Additionally, the targeted population will remain consistent, and there was more than a year between the test version of the survey and AETC's first administration of the revised instrument. Since this was a test survey, additional data should still be collected in order to have more confidence in the reliability of the prevalence estimates and properties of the scales and items on the surveys. However, the data do provide a nice initial baseline for AETC to build from in the future.

Basic Military Training Abuse and Misconduct Survey
Trainee Survey

Prepared by

The RAND Corporation

NOTE: This survey will be computer programmed and administered electronically, so the on-screen appearance of some questions will differ from that shown in this document (e.g., a question that includes a long list of items might be divided across two pages; individual questions in a sequence might appear on separate pages rather than the same page).

Notes about the survey programming appear in red.
Notes about the survey analysis appear in blue.

Estimated average time to complete the survey: 15–20 minutes. Some trainees may need additional time, depending on their experiences at BMT. We recommend providing trainees a total of 45 minutes to take the full survey.

Survey Administration Script

RAND recommends that the below instructions be delivered by a prerecorded video prior to the start of the survey, as well as appear at the beginning of the survey for trainees to review in written form. Thus, a significant portion of the content below is intentionally repeated in the subsequent "Instructions and Consent" section. Should there be technical glitches with showing the instructions video, the survey administrators should deliver the instructions orally before instructing trainees to open the survey. RAND recommends that civilians, rather than military personnel, administer the survey.

Note to BMT survey staff: Request that trainees be taken to the restrooms immediately prior to the survey to minimize disruptions during the survey.

Begin with introductions by the BMT survey staff, such as, "Hello, my name is [X] and I am the [list job title] here at BMT."

The purpose of the survey is to provide all trainees a <u>confidential</u> way to report to BMT leadership any incidents of bullying between trainees, maltreatment, maltraining, unprofessional relationships, and unwanted sexual experiences happening at BMT. The survey also explores why trainees may or may not report these types of experiences. Your responses will be combined with other trainees' responses, analyzed, and reported to BMT leaders so they can understand what is going on in the training environment and where there might be problems they need to address.

The survey is completely confidential. We are not asking for your name, an identification number, or your contact information. All trainees are asked to complete this survey at the end of their training.

Just in case your answers could potentially reveal who you are, we are taking additional steps to ensure confidentiality, such as:
- using software programs that allow us to restrict access to the survey data to just the BMT research staff
- reporting only survey results that have been grouped together so that the information will not give away your identity
- not disclosing individual survey responses, unless we face the exceptional situation of being required to by a judicial process.

Although you may have been required to attend this survey assembly, your participation in this survey is completely voluntary. This means you may skip any questions you do not wish to answer—even every question on the survey. If you decide to start the survey, you can still choose to stop taking it at any time.

There is no penalty if you choose not to participate. You will not be kept from graduating BMT or tech training or receive any type of punishment for not completing this survey. Participation will not help or harm your future assignments or promotions in the Air Force.

Some of the questions on unwanted sexual experiences may seem graphic and may cause discomfort to trainees asked to recall specific past experiences. However, you may skip any questions you do not wish to answer.

Your responses are important to helping Air Force leaders learn about abuse or misconduct at BMT so they can take additional steps to ensure a safe training environment for all trainees.

To avoid distracting others taking the survey, please come down to the front of the room if you need to:
- ask a question about the survey
- get technical help for any problems with your computer or the survey
- receive medical attention.

One of the staff will escort you outside of the classroom to ask you about your concerns.

Note to BMT survey staff: To prepare for any emergencies or urgent needs, have on hand telephone numbers for medical personnel, computing support, the chaplain on duty, the SARC, and Security Forces.

Although you will be required to stay for the entire 45-minute session, the survey should take less than the full 45 minutes to complete.

We want to emphasize to you that this is not an official channel for reporting abuse or misconduct. If you would like to talk to someone about any abuse or misconduct in order to get help or file an official report, you should contact one of the following directly: someone in your chain of command, a chaplain, a medical provider, the sexual assault response coordinator (SARC), Office of Special Investigations, or Security Forces. Also, you may leave a note in the critique boxes located throughout BMT.

When you have completed the survey, please remain in your seat. You may study or sit quietly, but please do not disturb others by talking.

Note: RAND recommends all trainees be required to remain in the survey room for a predetermined period of time (45 minutes will allow time for instructions and the trainees who may need a bit longer than average to complete the survey). Permitting trainees to leave as they complete their surveys would provide incentive for them to decline to participate or rush through the survey, and could suggest to peers which trainees are indicating abuse and misconduct and thus are receiving more of the follow-up questions. For those same reasons, trainees should not be told that they must all remain until the last participating trainee completes the survey.

Note to BMT survey staff: In the oral instructions, explain here how trainees should access the survey (e.g., "Please click on the survey icon on your computer desktop—you may now begin).

Regardless of whether you intend to fill out the questionnaire or not, please open the survey on your computer—you'll see the instructions you just heard and then will be able to indicate whether you do or do not wish to participate.

Instructions and Consent

[Programming note: The section title should be repeated on each screen of the survey for that section.]

The purpose of the survey is to provide all trainees a <u>confidential</u> way to report to BMT leadership any incidents of bullying between trainees, maltreatment, maltraining, unprofessional relationships, and unwanted sexual experiences happening at BMT. The survey also explores why trainees may or may not report these types of experiences. Your responses will be combined with other trainees' responses, analyzed, and reported to BMT leaders so they can understand what is going on in the training environment and where there might be problems they need to address.

The survey is completely confidential. We are not asking for your name, an identification number, or your contact information. All trainees are asked to complete this survey at the end of their training.

Just in case your answers could potentially reveal who you are, we are taking additional steps to ensure confidentiality, such as:

- using software programs that allow us to restrict access to the survey data to just the BMT research staff
- reporting only survey results that have been grouped together so that the information will not give away your identity
- not disclosing individual survey responses, unless we face the exceptional situation of being required to by a judicial process.

********** PAGE BREAK **********

Although you may have been required to attend this survey assembly, your participation in this survey is completely voluntary. This means you may skip any questions you do not wish to answer—even every question on the survey. If you decide to start the survey, you can still choose to stop taking it at any time.

There is no penalty if you choose not to participate. You will not be kept from graduating BMT or tech training or receive any type of punishment for not completing this survey. Participation will not help or harm your future assignments or promotions in the Air Force.

Some of the questions on unwanted sexual experiences may seem graphic and may cause discomfort to trainees asked to recall specific past experiences. However, you may skip any questions you do not wish to answer. Your responses are important to helping Air Force leaders learn about abuse or misconduct at BMT so they can take additional steps to ensure a safe training environment for all trainees.

To avoid distracting others taking the survey, please come down to the front of the room if you need to:

- ask a question about the survey
- get technical help for any problems with your computer or the survey
- receive medical attention.

One of the staff will escort you outside of the classroom to ask you about your concerns.

Although you will be required to stay for the entire 45-minute session, the survey should take less than the full 45 minutes to complete.

********** PAGE BREAK **********

We want to emphasize to you that this is not an official channel for reporting abuse or misconduct. If you would like to talk to someone about any abuse or misconduct in order to get help or file an official report, you should contact one of the following directly: someone in your chain of command, a chaplain, a medical provider, the sexual assault response coordinator (SARC), Office of Special Investigations, or Security Forces. Also, you may leave a note in the critique boxes located throughout BMT.

When you have completed the survey, please remain in your seat. You may study or sit quietly, but please do not disturb others by talking.

> **Please indicate whether you do or do not consent to participate in this study:**
>
> o **I have read the above statement about this study and volunteer to participate.**
>
> o **I don't want to participate in this study, but I don't want anyone to know that I am opting out. Please allow me to advance through the survey without filling any of it out.**
>
> o **I do not want to participate in this study, and I would like to exit the survey now.**

********** PAGE BREAK **********

Background

Before you begin the survey, we would like to ask you just one question about your background.

What is your gender?

- o   Male
- o   Female

********** PAGE BREAK **********

Programming note: Trainees should be allowed to skip any questions or even all questions on the survey. For all routing in the survey, a skipped question should be treated as if the trainee responded "no" or "never" to the question, unless otherwise indicated. For example, if they skipped all the "experience" questions in any domain, they should move forward in the survey as if they answered "never" to all of those questions.

Programming note: Trainees should receive an error message requesting a correction if they select mutually exclusive options on a "select all that apply" item (e.g., if trainees selected both a yes and no response, such as in item 1.2a).

Section I

Bullying: (1) Create a dichotomous variable indicating how many trainees experienced at least one item occurring versus no items occurring; (2) create an ordinal variable representing the greatest frequency that the trainee experienced any item (never = 1 to 5 = daily); (3) create a dichotomous variable for experience of each item; (4) create a summed scale by summing scores on items a–d and f–g.

1.1 Below is a list of things another <u>trainee</u> may have done while you were at BMT. For these questions, please don't consider anything that happened to other trainees. Think only about whether these things happened to you.

During BMT, did another trainee . . .

	Never	*Once or twice*	*A few times*	*Weekly*	*Daily*
a. Encourage other trainees to turn against you?	○	○	○	○	○
b. Try to embarrass you?	○	○	○	○	○
c. Try to get you into trouble with an MTI?	○	○	○	○	○
d. Steal something from you?	○	○	○	○	○
e. Please select "Daily" for this item to help us confirm that trainees are reading these items. [Screening item]	○	○	○	○	○
f. Threaten you?	○	○	○	○	○
g. Hit or kick you?	○	○	○	○	○

********** PAGE BREAK **********

IF A TRAINEE MARKED "NEVER" FOR ALL OF THE ABOVE ITEMS, GO TO QUESTION 1.2a BELOW ↓	IF A TRAINEE INDICATED THEY EXPERIENCED AT LEAST ONE OF THE ABOVE ITEMS, SKIP TO QUESTION 1.3 BELOW BEFORE PROCEEDING TO QUESTION 1.2b ↓
1.2a Are you aware of any trainees doing any of these things to <u>other</u> trainees while you were at BMT? (*Select all that apply*) ☐ No. **SKIP TO QUESTION 1.10.** ☐ Yes, I saw this happen. ☐ Yes, a trainee told me this happened to them. ☐ Yes, an MTI informed me of this happening. ☐ Yes, I heard about this happening from someone else. GO TO QUESTION 1.4. ********** PAGE BREAK **********	**1.3 On the previous page, you indicated that at least one of these situations happened to you personally. Did you tell any other trainees about any of these situations that happened to you?** ○ Yes, I told another trainee. ○ No, I did not tell another trainee. **1.2b Are you aware of any trainees doing any of these things to <u>other</u> trainees while you were at BMT?** (*Select all that apply*) [Note: Questions 1.2a and 1.2b are identical.] ☐ No. ☐ Yes, I saw this happen. ☐ Yes, a trainee told me this happened to them. ☐ Yes, an MTI informed me of this happening. ☐ Yes, I heard about this happening from someone else. GO TO QUESTION 1.4. ********** PAGE BREAK **********

1.4 For the situations in this section that happened to you personally, or that you were aware of happening to another trainee . . .

DID YOU REPORT ANY OF THE TRAINEE BEHAVIORS IN THIS SECTION TO ANY OF THE SOURCES BELOW? (*Please select all that apply, or indicate that you did not report any incidents.*)

- ☐ I told my dorm chief.
- ☐ I told an MTI.
- ☐ I told someone else in my chain of command.
- ☐ I told an officer or NCO outside of my chain of command.
- ☐ I told someone in Air Force law enforcement: Office of Special Investigations (OSI) or Security Forces (SF).
- ☐ I wrote it down on paper and put it in a critique box.
- ☐ I used the dorm hotline.
- ☐ I told a chaplain.
- ☐ I told the SARC (sexual assault response coordinator).
- ☐ **I did not report any incidents.** [Note: Skip to question 1.10 if no answer is provided for 1.4.]

********** PAGE BREAK **********

IF THEY SELECTED "I DID NOT REPORT ANY INCIDENTS," GO TO QUESTION 1.5 BELOW. ↓	IF THEY SELECTED THAT THEY REPORTED TO ANY SOURCE, GO TO QUESTION 1.6 BELOW. ↓

1.5 Please select the items below that describe why you did not report any incidents. (*Select all that apply. More options will be shown on the next page.*)

- ☐ I knew someone had already reported it.
- ☐ I thought someone else would report it.
- ☐ I only heard about it, so I wasn't sure if it was true.
- ☐ I didn't think there was anything wrong with it.
- ☐ I don't believe people should tell on one another.
- ☐ I didn't think anything would be done if I reported it.
- ☐ I didn't want anyone else to know it happened.
- ☐ I didn't think it was serious enough to report.
- ☐ I didn't think I would be believed.
- ☐ I was afraid reporting might cause trouble for my flight.

********** PAGE BREAK **********

Please select the items below that describe why you did not report any incidents. (*Select all that apply. Continued from the previous page.*)

- ☐ I handled it myself.
- ☐ I decided to put up with it.
- ☐ I didn't want the person who did it to get in trouble for it.
- ☐ I knew of others who were treated poorly for reporting.
- ☐ I was afraid the person who did it or their friends would try to get even with me for reporting.
- ☐ I was afraid trainees would punish me or mock me for reporting.

If you reported one incident, please answer the next questions about that one report. If you reported more than one incident, please think of the incident that you consider the most serious.

1.6 How seriously do you feel your report was taken?
- ○ Very seriously
- ○ Somewhat seriously
- ○ Not very seriously
- ○ Not at all seriously
- ○ I don't know

1.7 What happened with the behavior you complained about after you reported it?
- ○ The behavior didn't happen again.
- ○ The behavior continued or got worse.
- ○ I don't know: the behavior was happening to someone else.

1.8 What happened to you after the report? (*Select all that apply*)
- ☐ I got support to help me deal with what happened.
- ☐ The person I reported it to praised me for reporting.
- ☐ I got in trouble for my own misbehavior or infraction.
- ☐ The person who did it tried to get even with me for reporting.
- ☐ Trainees tried to get even with me for reporting.
- ☐ MTIs tried to get even with me for reporting.
- ☐ None of the above happened to me.

********** PAGE BREAK **********

93

- [ ] I was afraid MTIs would punish me, recycle me, or mock me for reporting.

- [ ] I didn't think my report would be kept confidential.

- [ ] I wanted to report it anonymously but didn't know a safe way to do that.

- [ ] I was afraid of getting into trouble for something I shouldn't have been doing.

- [ ] Other

GO TO THE NEXT PAGE (QUESTION 1.10).

********** PAGE BREAK **********

1.9 If you could do it over, would you still decide to report the incident?
- o Yes
- o No

GO TO THE NEXT PAGE (QUESTION 1.10).

********** PAGE BREAK **********

Climate for bullying (squadron leader actions): Average items a–d to form scale.

1.10 BMT Trainee Rules of Conduct state that trainees are required to act in a respectful, professional manner at all times, including when interacting with other trainees. This means that bullying behaviors are not acceptable. Examples of bullying include calling another trainee insulting names, hitting another trainee, and spreading lies about a trainee.

The following questions ask you about the extent to which rules against <u>bullying behaviors</u> are enforced at BMT. **For questions about squadron leaders, we are referring to those Air Force NCOs and officers with squadron-wide leadership responsibilities: the squadron commander, director of operations, superintendent, and first sergeant.** Please respond based on what you believe about your squadron leaders, even if you do not have direct knowledge about their attitudes or actions on this specific type of behavior.

	Strongly disagree	Disagree	Neither disagree nor agree	Agree	Strongly agree
a. Squadron leaders make honest efforts to stop bullying.	○	○	○	○	○
b. Squadron leaders encourage the reporting of bullying.	○	○	○	○	○
c. Squadron leaders take actions to prevent bullying.	○	○	○	○	○
d. Squadron leaders would correct or discipline a trainee who bullies another trainee.	○	○	○	○	○

********** PAGE BREAK **********

Section II

2.1 Below is a list of things MTIs may have done while you were at BMT. For these questions, please don't consider anything that happened to other trainees. Think only about whether these things happened to you.

During BMT, did an MTI . . .

	Never	Once or twice	A few times	Weekly	Daily
a. Discipline only you when others made the same mistakes?	o	o	o	o	o
b. Unfairly push you to quit or leave BMT?	o	o	o	o	o
c. Assign you activities unrelated to training objectives (for example, asked you to do his/her personal errands)?	o	o	o	o	o
d. Assign you training tasks that were against the rules (for example, required PT in the latrine)?	o	o	o	o	o
e. Encourage you to mistreat another trainee?	o	o	o	o	o
f. Make you do PT, drill, or outside work details in unsafe conditions (for example, during black flag conditions/temperatures above 90 degrees, extreme cold weather)?	o	o	o	o	o
g. Search your private mail or property for personal reasons?	o	o	o	o	o
h. Take a picture or videotape of you for personal reasons?	o	o	o	o	o
i. Deny you access to BMT services (for example, medical, SARC, chaplain)?	o	o	o	o	o

********* PAGE BREAK **********

96

During BMT, did an MTI . . .

	Never	Once or twice	A few times	Weekly	Daily
j. Deny you other rights provided by BMT (for example, withheld your mail or refused your authorized phone call)?	○	○	○	○	○
k. Please select "A few times" for this item to help us confirm that trainees are reading these items. [Screening item]	○	○	○	○	○
l. Call you insulting names (for example, "fatso," "ugly," or "idiot")?	○	○	○	○	○
m. Make negative comments about your race, ethnicity, religion, gender, or sexual orientation?	○	○	○	○	○
n. Threaten to hurt you?	○	○	○	○	○
o. Intentionally damage something of yours?	○	○	○	○	○
p. Hit or punch an object when angry (for example, a wall, window, table, or other object)?	○	○	○	○	○
q. Intentionally throw something at you?	○	○	○	○	○
r. Use physical force with you (for example, poked, hit, grabbed, or shoved you)?	○	○	○	○	○

********** PAGE BREAK **********

IF A TRAINEE MARKED "NEVER" FOR ALL OF THE ABOVE ITEMS, GO TO QUESTION 2.2a BELOW. ↓

IF A TRAINEE INDICATED THEY EXPERIENCED AT LEAST ONE OF THE ABOVE ITEMS, SKIP TO QUESTION 2.3 BELOW BEFORE PROCEEDING TO QUESTION 2.2b. ↓

2.2a Are you aware of any MTIs doing any of these things to other trainees while you were at BMT? (*Select all that apply*)

☐ No. **SKIP TO QUESTION 2.10.**
☐ Yes, I saw this happen.
☐ Yes, a trainee told me this happened to them.
☐ Yes, an MTI informed me of this happening.
☐ Yes, I heard about this happening from someone else.

GO TO QUESTION 2.4.

********** PAGE BREAK **********

2.3 On the previous page, you indicated that at least one of these situations happened to you personally. Did you tell any other trainees about any of these situations that happened to you?
o   Yes, I told another trainee.
o   No, I did not tell another trainee.

2.2b Are you aware of any MTIs doing any of these things to other trainees while you were at BMT? (*Select all that apply*)
[Note: Questions 2.2a and 2.2b are identical.]

☐ No.
☐ Yes, I saw this happen.
☐ Yes, a trainee told me this happened to them.
☐ Yes, an MTI informed me of this happening.
☐ Yes, I heard about this happening from someone else.

GO TO QUESTION 2.4.

********** PAGE BREAK **********

2.4 For the situations in this section that happened to you personally, or that you were aware of happening to another trainee . . .

DID YOU REPORT ANY OF THE MTI BEHAVIORS IN THIS SECTION TO ANY OF THE SOURCES BELOW? (*Please select all that apply, or indicate that you did not report any incidents.*)

- ☐ I told my dorm chief.
- ☐ I told an MTI.
- ☐ I told someone else in my chain of command.
- ☐ I told an officer or NCO outside of my chain of command.
- ☐ I told someone in Air Force law enforcement: Office of Special Investigations (OSI) or Security Forces (SF).
- ☐ I wrote it down on paper and put it in a critique box.
- ☐ I used the dorm hotline.
- ☐ I told a chaplain.
- ☐ I told the SARC (sexual assault response coordinator).
- ☐ **I did not report any incidents.** [Note: Skip to question 2.10 if no answer is provided for 2.4.]

********** PAGE BREAK **********

IF THEY SELECTED "I DID NOT REPORT ANY INCIDENTS," GO TO QUESTION 2.5 BELOW.	IF THEY SELECTED THAT THEY REPORTED TO ANY SOURCE, GO TO QUESTION 2.6 BELOW.

2.5 Please select the items below that describe why you did not report any incidents. (*Select all that apply. More options will be shown on the next page.*)

- ☐ I knew someone had already reported it.
- ☐ I thought someone else would report it.
- ☐ I only heard about it, so I wasn't sure if it was true.
- ☐ I didn't think there was anything wrong with it.
- ☐ I don't believe people should tell on one another.
- ☐ I didn't think anything would be done if I reported it.
- ☐ I didn't want anyone else to know it happened.
- ☐ I didn't think it was serious enough to report.
- ☐ I didn't think I would be believed.
- ☐ I was afraid reporting might cause trouble for my flight.

********** PAGE BREAK **********

Please select the items below that describe why you did not report any incidents. (*Select all that apply. Continued from the previous page.*)

- ☐ I handled it myself.
- ☐ I decided to put up with it.
- ☐ I didn't want the person who did it to get in trouble for it.
- ☐ I knew of others who were treated poorly for reporting.
- ☐ I was afraid the person who did it or their friends would try to get even with me for reporting.
- ☐ I was afraid trainees would punish me or mock me for reporting.

If you reported one incident, please answer the next questions about that one report. If you reported more than one incident, please think of the incident that you consider the most serious.

2.6 How seriously do you feel your report was taken?
- ○ Very seriously
- ○ Somewhat seriously
- ○ Not very seriously
- ○ Not at all seriously
- ○ I don't know

2.7 What happened with the behavior you complained about after you reported it?
- ○ The behavior didn't happen again.
- ○ The behavior continued or got worse.
- ○ I don't know: the behavior was happening to someone else.

2.8 What happened to you after the report? (*Select all that apply*)
- ☐ I got support to help me deal with what happened.
- ☐ The person I reported it to praised me for reporting.
- ☐ I got in trouble for my own misbehavior or infraction.
- ☐ The person who did it tried to get even with me for reporting.
- ☐ Trainees tried to get even with me for reporting.
- ☐ MTIs tried to get even with me for reporting.
- ☐ None of the above happened to me.

********** PAGE BREAK **********

99

- ☐ I was afraid MTIs would punish me, recycle me, or mock me for reporting.

- ☐ I didn't think my report would be kept confidential.

- ☐ I wanted to report it anonymously but didn't know a safe way to do that.

- ☐ I was afraid of getting into trouble for something I shouldn't have been doing.

- ☐ Other

GO TO THE NEXT PAGE (QUESTION 2.10).

********** PAGE BREAK **********

2.9 If you could do it over, would you still decide to report the incident?
- ○ Yes
- ○ No

GO TO THE NEXT PAGE (QUESTION 2.10).

********** PAGE BREAK **********

Climate for maltreatment/maltraining (squadron leader actions): Average items a–d to form scale.

2.10 BMT policies establish approved training methods and appropriate interactions between MTIs and trainees. MTIs making trainees perform humiliating tasks, physical exercise in unsafe conditions, threatening or hitting trainees, and using crude or offensive language are examples of policy violations that BMT calls maltreatment or maltraining.

The following questions ask you about the extent to which policies against <u>maltreatment and maltraining</u> are enforced at BMT. **For questions about squadron leaders, we are referring to those Air Force NCOs and officers with squadron-wide leadership responsibilities: the squadron commander, director of operations, superintendent, and first sergeant.** Please respond based on what you believe about your squadron leaders, even if you do not have direct knowledge about their attitudes or actions on this specific type of behavior.

	Strongly disagree	Disagree	Neither disagree nor agree	Agree	Strongly agree
a. Squadron leaders make honest efforts to stop maltreatment and maltraining.	○	○	○	○	○
b. Squadron leaders encourage the reporting of maltreatment and maltraining.	○	○	○	○	○
c. Squadron leaders take actions to prevent maltreatment and maltraining.	○	○	○	○	○
d. Squadron leaders would correct or discipline an MTI who engages in maltreatment or maltraining.	○	○	○	○	○

********** PAGE BREAK **********

Section III

3.1 Below is a list of some other things <u>MTIs</u> may have done while you were at BMT. For these questions, please don't consider anything that happened to other trainees. Think only about whether these things happened to you.

During BMT, did an MTI . . .

	Never	*Once or twice*	*A few times*	*Weekly*	*Daily*
a. Ask you to "just call me by my first name"?	○	○	○	○	○
b. Drink alcohol with you?	○	○	○	○	○
c. Flirt with you?	○	○	○	○	○
d. Give you more privileges than others even though you didn't earn them?	○	○	○	○	○
e. Contact you through non–Air Force channels for personal reasons (for example, by note, phone, email, Internet, or text)?	○	○	○	○	○
f. Share sexual jokes with you?	○	○	○	○	○
g. Meet you alone?	○	○	○	○	○
h. Talk about <u>his or her</u> sex life with you?	○	○	○	○	○
i. Talk about <u>your</u> sex life with you?	○	○	○	○	○

********** PAGE BREAK **********

During BMT, did an MTI . . .	Never	Once or twice	A few times	Weekly	Daily
j. Talk about dating you after you graduate?	o	o	o	o	o
k. Invite you to a social gathering (for example, parties or cookouts)?	o	o	o	o	o
l. Offer to give or loan you money or pay for something for you?	o	o	o	o	o
m. Ask you to give or loan them money or buy something?	o	o	o	o	o
n. Please select "Daily" for this item to help us confirm that trainees are reading these items. [Screening item]	o	o	o	o	o
o. Use your cell phone or other personal property?	o	o	o	o	o
p. Have a romantic relationship with you?	o	o	o	o	o
q. Engage in any type of sexual activity with you?	o	o	o	o	o

********** PAGE BREAK **********

IF A TRAINEE MARKED "NEVER" FOR ALL OF THE ABOVE ITEMS, GO TO QUESTION 3.2a BELOW. ↓	IF A TRAINEE INDICATED THEY EXPERIENCED AT LEAST ONE OF THE ABOVE ITEMS, SKIP TO QUESTION 3.3 BELOW BEFORE PROCEEDING TO QUESTION 3.2b. ↓

3.2a Are you aware of any MTIs doing any of these things to <u>other</u> trainees while you were at BMT?
(*Select all that apply*)

☐ No. **SKIP TO QUESTION 3.10.**
☐ Yes, I saw this happen.
☐ Yes, a trainee told me this happened to them.
☐ Yes, an MTI informed me of this happening.
☐ Yes, I heard about this happening from someone else.

GO TO QUESTION 3.4.

********** PAGE BREAK **********

3.3 On the previous page, you indicated that at least one of these situations happened to you personally. Did you tell any other trainees about any of these situations that happened to you?
○ Yes, I told another trainee.
○ No, I did not tell another trainee.

3.2b Are you aware of any MTIs doing any of these things to <u>other</u> trainees while you were at BMT?
(*Select all that apply*)
[Note: Questions 3.2a and 3.2b are identical.]

☐ No.
☐ Yes, I saw this happen.
☐ Yes, a trainee told me this happened to them.
☐ Yes, an MTI informed me of this happening.
☐ Yes, I heard about this happening from someone else.

GO TO QUESTION 3.4.

********** PAGE BREAK **********

3.4 For the situations in this section that happened to you personally, or that you were aware of happening to another trainee . . .

DID YOU REPORT ANY OF THE MTI BEHAVIORS IN THIS SECTION TO ANY OF THE SOURCES BELOW? (*Please select all that apply, or indicate that you did not report any incidents.*)

- ☐ I told my dorm chief.
- ☐ I told an MTI.
- ☐ I told someone else in my chain of command.
- ☐ I told an officer or NCO outside of my chain of command.
- ☐ I told someone in Air Force law enforcement: Office of Special Investigations (OSI) or Security Forces (SF).
- ☐ I wrote it down on paper and put it in a critique box.
- ☐ I used the dorm hotline.
- ☐ I told a chaplain.
- ☐ I told the SARC (sexual assault response coordinator).
- ☐ **I did not report any incidents.** [Note: Skip to question 3.10 if no answer is provided for 3.4.]

********** PAGE BREAK **********

IF THEY SELECTED "I DID NOT REPORT ANY INCIDENTS," GO TO QUESTION 3.5 BELOW.	IF THEY SELECTED THAT THEY REPORTED TO ANY SOURCE, GO TO QUESTION 3.6 BELOW.

3.5 Please select the items below that describe why you did not report any incidents. (*Select all that apply. More options will be shown on the next page.*)

- ☐ I knew someone had already reported it.
- ☐ I thought someone else would report it.
- ☐ I only heard about it, so I wasn't sure if it was true.
- ☐ I didn't think there was anything wrong with it.
- ☐ I don't believe people should tell on one another.
- ☐ I didn't think anything would be done if I reported it.
- ☐ I didn't want anyone else to know it happened.
- ☐ I didn't think it was serious enough to report.
- ☐ I didn't think I would be believed.
- ☐ I was afraid reporting might cause trouble for my flight.

********** PAGE BREAK **********

Please select the items below that describe why you did not report any incidents. (*Select all that apply. Continued from the previous page.*)

- ☐ I handled it myself.
- ☐ I decided to put up with it.
- ☐ I didn't want the person who did it to get in trouble for it.
- ☐ I knew of others who were treated poorly for reporting.
- ☐ I was afraid the person who did it or their friends would try to get even with me for reporting.
- ☐ I was afraid trainees would punish me or mock me for reporting.

If you reported one incident, please answer the next questions about that one report. If you reported more than one incident, please think of the incident that you consider the most serious.

3.6 How seriously do you feel your report was taken?
- ○ Very seriously
- ○ Somewhat seriously
- ○ Not very seriously
- ○ Not at all seriously
- ○ I don't know

3.7 What happened with the behavior you complained about after you reported it?
- ○ The behavior didn't happen again.
- ○ The behavior continued or got worse.
- ○ I don't know: the behavior was happening to someone else.

3.8 What happened to you after the report? (*Select all that apply.*)
- ☐ I got support to help me deal with what happened.
- ☐ The person I reported it to praised me for reporting.
- ☐ I got in trouble for my own misbehavior or infraction.
- ☐ The person who did it tried to get even with me for reporting.
- ☐ Trainees tried to get even with me for reporting.
- ☐ MTIs tried to get even with me for reporting.
- ☐ None of the above happened to me.

********** PAGE BREAK **********

- ☐ I was afraid MTIs would punish me, recycle me, or mock me for reporting.
- ☐ I didn't think my report would be kept confidential.
- ☐ I wanted to report it anonymously but didn't know a safe way to do that.
- ☐ I was afraid of getting into trouble for something I shouldn't have been doing.
- ☐ Other

GO TO THE NEXT PAGE (QUESTION 3.10).

********** PAGE BREAK **********

3.9 If you could do it over, would you still decide to report the incident?
- o Yes
- o No

GO TO THE NEXT PAGE (QUESTION 3.10).

********** PAGE BREAK **********

Climate for unprofessional relationships (squadron leader actions): Average items a–d to form scale.

3.10 BMT policy states that MTIs are not allowed to develop friendships or romantic relationships with trainees or show favoritism to specific trainees. The Air Force deems these unprofessional relationships, even if they develop only through cards, letters, emails, phone calls, the Internet, or instant messaging. Examples of behaviors that violate Air Force professional-relationship policies include MTIs' giving individual trainees special privileges as well as MTIs dating, drinking alcohol, or sharing sexual stories with trainees.

The following questions ask you about the extent to which policies against <u>unprofessional relationships</u> are enforced at BMT. **For questions about squadron leaders, we are referring to those Air Force NCOs and officers with squadron-wide leadership responsibilities: the squadron commander, director of operations, superintendent, and first sergeant.** Please respond based on what you believe about your squadron leaders, even if you do not have direct knowledge about their attitudes or actions on this specific type of behavior.

	Strongly disagree	Disagree	Neither disagree nor agree	Agree	Strongly agree
a. Squadron leaders make honest efforts to stop unprofessional relationships.	○	○	○	○	○
b. Squadron leaders encourage the reporting of unprofessional relationships.	○	○	○	○	○
c. Squadron leaders take actions to prevent unprofessional relationships.	○	○	○	○	○
d. Squadron leaders would correct or discipline an MTI who engages in an unprofessional relationship.	○	○	○	○	○

********** PAGE BREAK **********

107

Section IV

Sexual harassment: (1) Create a dichotomous variable indicating how many trainees experienced at least one item occurring versus no items occurring; (2) create an ordinal variable representing the greatest frequency that the trainee experienced any item (never = 1 to 5 = daily); (3) create a dichotomous variable for whether a trainee experienced any item in the following groupings:

- Sexist hostility (a, b, c)
- Challenges to masculinity/femininity (d, e)
- Sexual hostility (f, g, h, i)
- Sexual coercion (j, l, m, n)
- Unwanted sexual attention (o, p, q).

4.1 Below is a list of things <u>someone</u> may have done to you while you were at BMT (for example, MTIs or other trainees). In these questions you are asked about sexual or gender-related talk and behavior that was unwanted, uninvited, and in which you did not participate willingly. Please don't consider anything that happened to other trainees. Think only about whether these things happened to you.

<u>During BMT, has anyone, male or female, . . .</u>

		Never	*Once or twice*	*A few times*	*Weekly*	*Daily*
a.	Treated you "differently" because of your gender (for example, mistreated, slighted, or ignored you)?	o	o	o	o	o
b.	Displayed, used, or distributed sexist or suggestive materials (for example, pictures, stories, or pornography which you found offensive)?	o	o	o	o	o
c.	Made offensive sexist remarks (for example, suggesting that people of your gender are not suited for the kind of work you do)?	o	o	o	o	o
d.	Called you gay as an insult (for example, "fag," "queer," or "dyke")?	o	o	o	o	o
e.	Insulted you by saying you were not acting like a real man or real woman (for example, called you a "sissy" or said you were "acting like a girl" or "pretending to be a man")?	o	o	o	o	o
f.	Repeatedly told sexual stories or jokes that were offensive to you?	o	o	o	o	o
g.	Made unwelcome attempts to draw you into a discussion of sexual matters (for example, attempted to discuss or comment on your sex life)?	o	o	o	o	o
h.	Made gestures or used body language of a sexual nature which embarrassed or offended you?	o	o	o	o	o
i.	Made offensive remarks about your appearance, body, or sexual activities?	o	o	o	o	o

During BMT, has anyone, male or female, . . .		Never	Once or twice	A few times	Weekly	Daily
j.	Made you feel like you were being bribed with some sort of reward or special treatment to engage in sexual behavior?	O	O	O	O	O
k.	Please select "Weekly" for this item to help us confirm that trainees are reading these items. **[Screening item]**	O	O	O	O	O
l.	Made you feel threatened with some sort of retaliation for not being sexually cooperative (for example, by mentioning an upcoming test)?	O	O	O	O	O
m.	Treated you badly for refusing to have sex with him or her?	O	O	O	O	O
n.	Implied you would receive better performance evaluations or better treatment if you were sexually cooperative?	O	O	O	O	O
o.	Made unwanted attempts to establish a sexual relationship with you despite your efforts to discourage it?	O	O	O	O	O
p.	Touched you in a way that made you feel uncomfortable?	O	O	O	O	O
q.	Made unwanted attempts to touch or kiss you?	O	O	O	O	O

********** PAGE BREAK **********

4.2a Are you aware of any of these things happening to other trainees while you were at BMT? (*Select all that apply.*)

☐   No. SKIP TO QUESTION 4.13.
☐   Yes, I saw this happen.
☐   Yes, a trainee told me this happened to them.
☐   Yes, an MTI informed me of this happening.
☐   Yes, I heard about this happening from someone else.

GO TO NEXT PAGE (QUESTION 4.7).

********** PAGE BREAK **********

4.3 On the previous pages, you indicated that at least one of these situations happened to you personally. Did you tell any other trainees about any of these situations that happened to you?

o   Yes, I told another trainee.
o   No, I did not tell another trainee.

********** PAGE BREAK **********

Think about the situation(s) you selected in this section as happening to you while you were at BMT. Now pick the behavior that you consider to be the most serious or that had the greatest effect on you. For the next questions, please think of that worst behavior.

4.4 How many people did this to you? (*Mark one.*)
o   One person
o   More than one person

4.5 Were they . . . ? (*Mark one.*)
o   Male(s)
o   Female(s)
o   Both male(s) and female(s)

4.6 Were they a/an . . . ? (*Select all that apply.*)
☐   Trainee
☐   MTI
☐   Other military personnel
☐   Non-military personnel

********** PAGE BREAK **********

4.2b The previous questions asked whether any of these situations happened to you personally. Are you aware of any of these things happening to other trainees while you were at BMT? (*Select all that apply.*)
[Note: Questions 4.2a and 4.2b are identical.]

☐   No.
☐   Yes, I saw this happen.
☐   Yes, a trainee told me this happened to them.
☐   Yes, an MTI informed me of this happening.
☐   Yes, I heard about this happening from someone else.

GO TO NEXT PAGE (QUESTION 4.7).

********** PAGE BREAK **********

4.7 For the situations in this section that happened to you personally, or that you were aware of happening to another trainee, . . .

DID YOU REPORT ANY OF THE BEHAVIORS IN THIS SECTION TO ANY OF THE SOURCES BELOW? (*Please select all that apply, or indicate that you did not report any incidents.*)

- ☐ I told my dorm chief.
- ☐ I told an MTI.
- ☐ I told someone else in my chain of command.
- ☐ I told an officer or NCO outside of my chain of command.
- ☐ I told someone in Air Force law enforcement: Office of Special Investigations (OSI) or Security Forces (SF).
- ☐ I wrote it down on paper and put it in a critique box.
- ☐ I used the dorm hotline.
- ☐ I told a chaplain.
- ☐ I told the SARC (sexual assault response coordinator).
- ☐ **I did not report any incidents.** [Note: Skip to question 4.13 if no answer is provided for 4.7.]

********** PAGE BREAK **********

IF THEY SELECTED "I DID NOT REPORT ANY INCIDENTS," GO TO QUESTION 4.8 BELOW. ↓	IF THEY SELECTED THAT THEY REPORTED TO ANY SOURCE, GO TO QUESTION 4.9 BELOW. ↓

4.8 Please select the items below that describe why you did not report any incidents. (*Select all that apply. More options will be shown on the next page.*)

- ☐ I knew someone had already reported it.
- ☐ I thought someone else would report it.
- ☐ I only heard about it, so I wasn't sure if it was true.
- ☐ I didn't think there was anything wrong with it.
- ☐ I don't believe people should tell on one another.
- ☐ I didn't think anything would be done if I reported it.
- ☐ I didn't want anyone else to know it happened.
- ☐ I didn't think it was serious enough to report.
- ☐ I didn't think I would be believed.
- ☐ I was afraid reporting might cause trouble for my flight.

********** PAGE BREAK **********

Please select the items below that describe why you did not report any incidents. (*Select all that apply. Continued from the previous page.*)

- ☐ I handled it myself.
- ☐ I decided to put up with it.
- ☐ I didn't want the person who did it to get in trouble for it.
- ☐ I knew of others who were treated poorly for reporting.
- ☐ I was afraid the person who did it or their friends would try to get even with me for reporting.
- ☐ I was afraid trainees would punish me or mock me for reporting.

If you reported one incident, please answer the next questions about that one report. If you reported more than one incident, please think of the incident that you consider the most serious.

4.9 How seriously do you feel your report was taken?
- ○ Very seriously
- ○ Somewhat seriously
- ○ Not very seriously
- ○ Not at all seriously
- ○ I don't know

4.10 What happened with the behavior you complained about after you reported it?
- ○ The behavior didn't happen again.
- ○ The behavior continued or got worse.
- ○ I don't know: the behavior was happening to someone else.

4. 11 What happened to you after the report? (*Select all that apply.*)
- ☐ I got support to help me deal with what happened.
- ☐ The person I reported it to praised me for reporting.
- ☐ I got in trouble for my own misbehavior or infraction.
- ☐ The person who did it tried to get even with me for reporting.
- ☐ Trainees tried to get even with me for reporting.
- ☐ MTIs tried to get even with me for reporting.
- ☐ None of the above happened to me.

********** PAGE BREAK **********

111

□ I was afraid MTIs would punish me, recycle me, or mock me for reporting.

□ I didn't think my report would be kept confidential.

□ I wanted to report it anonymously but didn't know a safe way to do that.

□ I was afraid of getting into trouble for something I shouldn't have been doing.

□ Other

GO TO THE NEXT PAGE (QUESTION 4.13).

********** PAGE BREAK **********

4.12 If you could do it over, would you still decide to report the incident?
  o Yes.
  o No.

GO TO THE NEXT PAGE (QUESTION 4.13).

********** PAGE BREAK **********

Climate for sexual harassment (squadron leader actions): Average items a–d to form scale.

4.13 Air Force policy states: "Unwelcome sexual advances, requests for sexual favors, and other verbal or physical conduct of a sexual nature constitute sexual harassment when (1) submission to such conduct is made either explicitly or implicitly a term or condition of an individual's employment, (2) submission to or rejection of such conduct by an individual is used as the basis for employment decisions affecting such individual, or (3) such conduct has the purpose or effect of unreasonably interfering with an individual's work performance or creating an intimidating, hostile, or offensive working environment."

The following questions ask you about the extent to which these <u>sexual harassment</u> policies are enforced at BMT. **For questions about squadron leaders, we are referring to those Air Force NCOs and officers with squadron-wide leadership responsibilities: the squadron commander, director of operations, superintendent, and first sergeant.** Please respond based on what you believe about your squadron leaders, even if you do not have direct knowledge about their attitudes or actions on this specific type of behavior.

	Strongly disagree	Disagree	Neither disagree nor agree	Agree	Strongly agree
a. Squadron leaders make honest efforts to stop sexual harassment.	○	○	○	○	○
b. Squadron leaders encourage the reporting of sexual harassment.	○	○	○	○	○
c. Squadron leaders take actions to prevent sexual harassment.	○	○	○	○	○
d. Squadron leaders would correct or discipline someone who engages in sexual harassment.	○	○	○	○	○

********** PAGE BREAK **********

Section V

Unwanted sexual experiences: (1) Create a dichotomous variable indicating how many trainees experienced at least one item occurring versus no items occurring; (2) create a dichotomous variable for whether a trainee experienced any item in the following groupings:
- exposure (a)
- sexual contact (b)
- attempted rape (d, g, i)
- completed rape (c, f, h).

5.1 The following questions ask about unwanted sexual experiences that may have happened to you at BMT. Although unwanted sexual experiences can happen at any point in a person's life, **these questions refer to things that may have happened to you <u>while you were a BMT trainee</u>.** Unwanted sexual experiences can happen to men or women and be carried out by either a man or woman.

Some of the questions may seem graphic to you, but using correct terms is the best way to determine whether or not trainees have had these experiences. All the information you share will be confidential so your responses cannot be linked back to you personally.

a.	During BMT, did anyone show you private areas of their body or make you show them private areas of your body when you didn't want to? (By private areas we mean vagina or penis, anus, breast, inner thigh, and buttocks.)	o No o Yes
b.	During BMT, did anyone touch, kiss, or rub up against the private areas of your body when you didn't want them to?	o No o Yes
c.	During BMT, did anyone have oral sex with you when you didn't want to? (Please consider it oral sex anytime someone put their mouth on your vagina or penis or made you put your mouth on their vagina or penis [even if ejaculation did not occur]).	o No o Yes
d.	During BMT, did anyone TRY but fail to have oral sex with you when you didn't want to?	o No o Yes
e.	Please select "Yes" for this item to help us confirm that trainees are reading these items. [Screening item]	o No o Yes

[Programming note: Items f and g should be included only for female respondents, and should appear on the same pages as these other items. If respondents did not previously indicate gender, display f and g along with the "N/A I am a male" option.]

f.	During BMT, did anyone insert fingers, objects, or their penis into your vagina when you didn't want them to?	o No o Yes o N/A I am a male *[Note: Include this third option only when these items are being shown to a trainee who did not previously disclose gender.]*

114

g.	During BMT, did anyone TRY but fail to insert fingers, objects, or their penis into your vagina when you didn't want them to?	o No o Yes o N/A I am a male [*Note: Include this third option only when these items are being shown to a trainee who did not previously disclose gender.*]
h.	During BMT, did anyone insert fingers, objects, or their penis into your anus when you didn't want them to?	o No o Yes
i.	During BMT, did anyone TRY but fail to insert fingers, objects, or their penis into your anus when you didn't want them to?	o No o Yes

********** PAGE BREAK **********

115

5.2a Are you aware of any of these things happening to other trainees while you were at BMT? (Select all that apply.)

☐ No. **SKIP TO QUESTION 5.17.**
☐ Yes, I saw this happen.
☐ Yes, a trainee told me this happened to them.
☐ Yes, an MTI informed me of this happening.
☐ Yes, I heard about this happening from someone else.

GO TO NEXT PAGE (QUESTION 5.11).

********** PAGE BREAK **********

5.3 Did you tell any other trainees about any of these situations that happened to you?

o   Yes, I told another trainee.
o   No, I did not tell another trainee.

5.4 DURING BMT, how many different times did you have an unwanted sexual experience involving the exposure, touching, or attempted touching of private areas of the body? [*INSERT DROP DOWN BOX with 1, 2, 3, 4, 5, 6, 7, 8, 9, 10, 10+.*]

If this happened to you one time, please answer the next questions about that one incident. If this happened to you more than one time at BMT, please think of the most serious event or the one that had the greatest effect on you. [*PROGRAMMING NOTE: Repeat this instruction on all screens for items 5.5–5.9.*]

5.5 What did the person do to make you have an unwanted sexual experience without your consent? (Select all that apply.)
☐ Took me by surprise
☐ Overwhelmed me with constant arguments and pressure
☐ Threatened to end the relationship or spread rumors about me
☐ Showed displeasure, criticized my sexuality or attractiveness, or became angry
☐ Used their position of authority (MTI, supervisor) to make me
☐ Ignored me after I said or did something that showed that I did not agree
☐ Used some degree of physical force (twisted my arm, held me down) to make me
☐ Threatened to physically harm me or someone close to me.
☐ Threatened me with a weapon
☐ Did it when I was asleep, too drunk, drugged, or too out of it to stop what was happening
☐ Other

********** PAGE BREAK **********

5.6 Where did it occur?
o   In the dorms (including bays, latrines, closets, day rooms, flight offices)
o   In another BMT room or building (including classroom, chow hall, chapel)
o   At the BEAST training area
o   At another outdoor location (such as behind a building)
o   In a car, truck, van, or other vehicle
o   Somewhere else

5.7 How many people did this to you (during this one event)?
o   One person
o   More than one person
o   Don't know

5.8 Were they . . . ?
- o   Male(s)
- o   Female(s)
- o   Both male(s) and female(s)
- o   Don't know

5.9 Were they a(n) . . . ? (*Select all that apply.*)

☐ Trainee		☐ Non-military personnel
☐ MTI		☐ Don't know
☐ Other military personnel		

********** PAGE BREAK **********

5.10 Did you receive any of the following services for ANY unwanted sexual experience that occurred while you were at BMT? (*Select all that apply.*)
- ☐ Help from the SARC (Sexual Assault Response Coordinator)
- ☐ Victim advocacy services (for example, someone to accompany you to appointments and explain what to expect)
- ☐ Counseling services
- ☐ Support from a chaplain
- ☐ Medical services
- ☐ Legal services
- ☐ Help making arrangements so you didn't have to come in contact with anyone who did it to you

5.2b The previous questions asked whether any of these situations happened to you personally. Are you aware of any of these things happening to other trainees while you were at BMT?
(Select all that apply.)
[Note: Questions 5.2a and 5.2b are identical.]

- ☐ No.
- ☐ Yes, I saw this happen.
- ☐ Yes, a trainee told me this happened to them.
- ☐ Yes, an MTI informed me of this happening.
- ☐ Yes, I heard about this happening from someone else.

GO TO NEXT PAGE (QUESTION 5.11).

********** PAGE BREAK **********

117

5.11 For the situations in this section that happened to you personally, or that you were aware of happening to another trainee, . . .

DID YOU REPORT ANY OF THE UNWANTED SEXUAL EXPERIENCES IN THIS SECTION TO ANY OF THE SOURCES BELOW? (*Please select all that apply, or indicate that you did not report any incidents.*)

- ☐ I told my dorm chief.
- ☐ I told an MTI.
- ☐ I told someone else in my chain of command.
- ☐ I told an officer or NCO outside of my chain of command.
- ☐ I told someone in Air Force law enforcement: Office of Special Investigations (OSI) or Security Forces (SF).
- ☐ I wrote it down on paper and put it in a critique box.
- ☐ I used the dorm hotline.
- ☐ I told a chaplain.
- ☐ I told the SARC (sexual assault response coordinator).
- ☐ **I did not report any incidents.** [Note: Skip to question 5.17 if no answer is provided for 5.11.]

********** PAGE BREAK **********

IF THEY SELECTED "I DID NOT REPORT ANY INCIDENTS," GO TO QUESTION 5.12 BELOW. ↓	IF THEY SELECTED THAT THEY REPORTED TO ANY SOURCE, GO TO QUESTION 5.13 BELOW. ↓

5.12 Please select the items below that describe why you did not report any incidents. (*Select all that apply. More options will be shown on the next page.*)

- ☐ I knew someone had already reported it.
- ☐ I thought someone else would report it.
- ☐ I only heard about it, so I wasn't sure if it was true.
- ☐ I didn't think there was anything wrong with it.
- ☐ I don't believe people should tell on one another.
- ☐ I didn't think anything would be done if I reported it.
- ☐ I didn't want anyone else to know it happened.
- ☐ I didn't think it was serious enough to report.
- ☐ I didn't think I would be believed.
- ☐ I was afraid reporting might cause trouble for my flight.

********** PAGE BREAK **********

Please select the items below that describe why you did not report any incidents. (*Select all that apply. Continued from the previous page.*)

- ☐ I handled it myself.
- ☐ I decided to put up with it.
- ☐ I didn't want the person who did it to get in trouble for it.
- ☐ I knew of others who were treated poorly for reporting.
- ☐ I was afraid the person who did it or their friends would try to get even with me for reporting.
- ☐ I was afraid trainees would punish me or mock me for reporting.

If you reported one incident, please answer the next questions about that one report. If you reported more than one incident, please think of the incident that you consider the most serious.

5.13 How seriously do you feel your report was taken?

- ○ Very seriously
- ○ Somewhat seriously
- ○ Not very seriously
- ○ Not at all seriously
- ○ I don't know

5.14 What happened with the behavior you complained about after you reported it?

- ○ The behavior didn't happen again.
- ○ The behavior continued or got worse.
- ○ I don't know: the behavior was happening to someone else.

5.15 What happened to you after the report? (*Select all that apply.*)

- ☐ I got support to help me deal with what happened.
- ☐ The person I reported it to praised me for reporting.
- ☐ I got in trouble for my own misbehavior or infraction.
- ☐ The person who did it tried to get even with me for reporting.
- ☐ Trainees tried to get even with me for reporting.
- ☐ MTIs tried to get even with me for reporting.
- ☐ None of the above happened to me.

********** PAGE BREAK **********

☐ I was afraid MTIs would punish me, recycle, me or mock me for reporting.

☐ I didn't think my report would be kept confidential.

☐ I wanted to report it anonymously but didn't know a safe way to do that.

☐ I was afraid of getting into trouble for something I shouldn't have been doing.

☐ Other

GO TO THE NEXT PAGE (QUESTION 5.17).

********** PAGE BREAK **********

5.16 If you could do it over, would you still decide to report the incident?

○ Yes

○ No

GO TO THE NEXT PAGE (QUESTION 5.17).

********** PAGE BREAK **********

Climate for sexual assault (squadron leader actions): Average items a–d to form scale.

5.17 The Uniform Code of Military Justice (UCMJ) criminalizes various forms of unwanted sexual activity, including rape, sexual assault, and unwanted sexual contact. For the next sections, we use the term *sexual assault* to refer to all of these forms of unwanted sexual contact characterized by force, threats, intimidation, or abuse of authority, or when the victim does not or cannot consent to that sexual contact.

The following questions ask you about the extent to which military laws and policies on <u>sexual assault</u> are enforced at BMT. **For questions about squadron leaders, we are referring to those Air Force NCOs and officers with squadron-wide leadership responsibilities: the squadron commander, director of operations, superintendent, and first sergeant.** Please respond based on what you believe about your squadron leaders, even if you do not have direct knowledge about their attitudes or actions on this specific type of behavior.

	Strongly disagree	Disagree	Neither disagree nor agree	Agree	Strongly agree
a. Squadron leaders make honest efforts to stop sexual assault.	○	○	○	○	○
b. Squadron leaders encourage the reporting of sexual assault.	○	○	○	○	○
c. Squadron leaders take actions to prevent sexual assault.	○	○	○	○	○
d. Squadron leaders would discipline someone who engages in sexual assault.	○	○	○	○	○

********** PAGE BREAK **********

Section VI

The questions in this section now ask you about BMT Feedback and Support Systems.

6.1 Below is a list of people you may or may not see every day. How easy would it be for you to arrange to speak personally with the following people if you wanted to talk to them about problems at BMT like the ones mentioned in this survey?

	Very easy	Easy	Neither easy nor difficult	Difficult	Very difficult	Doesn't apply: we don't currently have one
a. Your instructor supervisor	○	○	○	○	○	○
b. Your flight commander	○	○	○	○	○	○
c. Your squadron superintendent	○	○	○	○	○	○
d. Your first sergeant	○	○	○	○	○	○
e. Your director of operations	○	○	○	○	○	○
f. Your squadron commander	○	○	○	○	○	○
g. One of your MTIs	○	○	○	○	○	
h. A chaplain	○	○	○	○	○	
i. A SARC (Sexual Assault Response Coordinator)	○	○	○	○	○	
j. A BMT doctor or nurse	○	○	○	○	○	
k. A BMT counselor or mental health professional	○	○	○	○	○	
l. AF law enforcement: Office of Special Investigations (OSI) or Security Forces (SF)	○	○	○	○	○	

********** PAGE BREAK **********

6.2 Which of the following people would you be able to recognize by sight if they walked past you? (*Select all that apply.*) [*Programing note: Omit any leader from this list that the trainee indicated in 6.1 that they do not currently have.*]

- ☐ My instructor supervisor
- ☐ My flight commander
- ☐ My squadron superintendent
- ☐ My first sergeant
- ☐ My director of operations
- ☐ My squadron commander
- ☐ A chaplain
- ☐ The SARC (Sexual Assault Response Coordinator)

********** PAGE BREAK **********

121

6.3 Please indicate the extent to which you agree or disagree with the following statements.

	Strongly disagree	Disagree	Neither disagree nor agree	Agree	Strongly agree
a. I could arrange to meet with the SARC without anyone in my flight knowing.	O	O	O	O	O
b. Sometimes it's just not possible to have a wingman with you wherever you go.	O	O	O	O	O
c. I would be willing to use the dorm hotline to report a problem at BMT or to ask for help.	O	O	O	O	O
d. BMT makes it easy to use a dorm hotline.	O	O	O	O	O
e. I would be able to report a problem on the dorm hotline without anyone knowing.	O	O	O	O	O
f. MTIs or others in the chain of command discourage trainees from using the dorm hotline to report misconduct.	O	O	O	O	O

********** PAGE BREAK **********

	Strongly disagree	Disagree	Neither disagree nor agree	Agree	Strongly agree
g. I would be willing to use a critique box to report a problem at BMT or to ask for help.	O	O	O	O	O
h. BMT makes it easy to use the critique boxes.	O	O	O	O	O
i. I would be able to put a comment in a critique box without someone noticing.	O	O	O	O	O
j. MTIs or others in the chain of command discourage trainees from using critique boxes to report misconduct.	O	O	O	O	O
k. If I experienced abuse or mistreatment from an MTI, there is at least one person at BMT in the chain of command I feel I could turn to for help (for example, a team chief, instructor supervisor, or first sergeant).	O	O	O	O	O
l. If I experienced abuse or mistreatment from an MTI, there is at least one person at BMT outside of the chain of command I feel I could turn to for help (for example, a chaplain, SARC, or a doctor).	O	O	O	O	O

********** PAGE BREAK **********

Closing Question

How open and honest did you feel you could be when answering these survey questions?

- ○ Not at all open or honest
- ○ Somewhat open and honest
- ○ Completely open and honest

********** PAGE BREAK **********

THANK YOU FOR PARTICIPATING IN THIS SURVEY

If you have experienced any of the situations you were asked about as part of this survey and need help or would like to make an official report, there are several options for you to do so:

- ✓ You can talk to someone in your chain of command (MTI, instructor supervisor, squadron commander)

- ✓ You can contact AF law enforcement (Office of Special Investigations or Security Forces)

- ✓ You can submit a complaint or report in one of the critique boxes located throughout BMT

- ✓ You can talk to your chaplain

- ✓ You can contact a SARC (sexual assault response coordinator) at 671-SARC (24-hour confidential hotline) if you were sexually assaulted or know someone at BMT who was sexually assaulted

- ✓ You can use one of the dorm hotline phones to contact the SARC or chaplain

The chaplain and SARC can provide CONFIDENTIAL help following a sexual assault if you have not made an official report and do not want them to tell anyone else what happened.

Basic Military Training MTI Survey

Prepared by

The RAND Corporation

NOTE: This survey will be computer programmed and administered electronically, so the on-screen appearance of some questions will differ from that shown in this document (e.g., a question that includes a long list of items might be divided across two pages; individual questions in a sequence might appear on separate pages rather than the same page).

Notes about the survey programming appear in red.
Notes about the survey analysis appear in blue.

Estimated average time to complete the survey: 25 minutes (excludes time for responding to open-ended questions). We recommend providing MTIs a total of 45 minutes to take the full survey.

Survey Administration Script

RAND recommends that these instructions be delivered by a prerecorded video prior to the start of the survey, as well as appear at the beginning of the survey for MTIs to review in written form. Thus, a significant portion of the content below is intentionally repeated in the subsequent "Instructions and Consent" section. Should there be technical glitches with showing the instructions video, the survey administrators should deliver the instructions orally before instructing MTIs to open the survey. RAND recommends that civilians, rather than military personnel, administer the survey.

Begin with introductions by the BMT survey staff, such as, "Hello, my name is [X] and I am the [list job title] here at BMT."

The purpose of this survey is to foster an effective Basic Military Training (BMT) environment that is safe for both trainees and MTIs and is free of abuse and misconduct. This survey will be given to MTIs every six months to a year during their tour to gather MTI feedback on their work experiences, quality of life, and the BMT environment.

All MTIs who have served as an MTI for at least one month have been asked to participate. The survey team will combine your responses with other MTIs' responses, analyze them, and report them to BMT leaders so they can better understand the issues MTIs face in the training environment and where there might be problems they need to address.

The survey is completely anonymous. We are not asking for your name, an identification number, or your contact information.

Although you may have been required to attend this survey session, your participation in this survey is completely voluntary. This means you may skip any questions you do not wish to answer—even every question on the survey. If you decide to start the survey, you can still choose to stop taking it at any time.

There is no penalty if you choose not to participate. Participation will not help or harm your future assignments or promotions in the Air Force.

Some of the questions on unwanted sexual experiences may seem graphic and may cause discomfort. However, you may skip any questions you do not wish to answer. Your responses are important for helping Air Force leaders learn about MTI working conditions as well as abuse or misconduct at BMT so they can take additional steps to ensure a positive and safe training environment.

The estimated average time to complete this survey is 25 minutes.

Note to BMT survey staff: In the oral instructions, explain here how MTIs should access the survey (e.g., "Please click on the survey icon on your computer desktop—you may now begin").

Regardless of whether you intend to fill out the questionnaire or not, please open the survey on your computer—you'll see the instructions you just heard and then will be able to indicate whether you do or do not wish to participate.

Instructions and Consent

[Programming note: The section title should be repeated on each screen of the survey for that section.]

The purpose of this survey is to foster an effective Basic Military Training (BMT) environment that is safe for both trainees and MTIs, and is free of abuse and misconduct. This survey will be given to MTIs every six months to a year during their tour to gather MTI feedback on their work experiences, quality of life, and the BMT environment.

All MTIs who have served as an MTI for at least one month have been asked to participate. The survey team will combine your responses with other MTIs' responses, analyze them, and report them to BMT leaders so they can better understand the issues MTIs face in the training environment and where there might be problems they need to address.

The survey is completely anonymous. We are not asking for your name, an identification number, or your contact information.

********** PAGE BREAK **********

Although you may have been required to attend this survey session, your participation in this survey is completely voluntary. This means you may skip any questions you do not wish to answer—even every question on the survey. If you decide to start the survey, you can still choose to stop taking it at any time.

There is no penalty if you choose not to participate. Participation will not help or harm your future assignments or promotions in the Air Force.

Some of the questions on unwanted sexual experiences may seem graphic and may cause discomfort. However, you may skip any questions you do not wish to answer. Your responses are important for helping Air Force leaders learn about MTI working conditions as well as abuse or misconduct at BMT so they can take additional steps to ensure a positive and safe training environment.

The estimated average time to complete this survey is 25 minutes.

Please indicate whether you do or do not consent to participate in this study:

o **I have read the above statement about this study and volunteer to participate.**

o **I do not want to participate in this study.**

********** PAGE BREAK **********

Background

Before you begin the survey, we would like to ask you the following two questions about your background.

How long have you been an MTI, in total?

- o 6 months or less
- o Greater than 6 months, but less than 2 years
- o 2 years or more

What is your primary duty?

- o Line
- o Supervisor
- o Other (WS/MS, BEAST Cadre, etc.)

********** PAGE BREAK **********

Section I

1.1 The following questions ask you about your attitudes toward being an MTI at BMT. Please read each statement carefully and indicate the extent to which you agree or disagree with the statement.

	Strongly disagree	Disagree	Neither disagree nor agree	Agree	Strongly agree
a. I enjoy discussing BMT with people outside it.	○	○	○	○	○
b. I do not feel like "part of the team" at BMT. [Reverse scored item]	○	○	○	○	○
c. BMT has a great deal of personal meaning for me.	○	○	○	○	○
d. I would be happy to extend my duty at BMT.	○	○	○	○	○
e. I would recommend becoming an MTI to others.	○	○	○	○	○

Job satisfaction (Highhouse and Becker, 1993): single item.

	Very dissatisfied	Dissatisfied	Neither dissatisfied nor satisfied	Satisfied	Very satisfied
1.2 Considering everything, how would you rate your overall satisfaction with your job as an MTI?	○	○	○	○	○

********** PAGE BREAK **********

Section II

2.1 The following questions ask you about the work environment at BMT. Please respond based on your experience over the last six months. For this section and all other sections that ask you about the last six months, if you have been an MTI for less time, please just answer based on the time you have been here.

	Strongly disagree	Disagree	Neither disagree nor agree	Agree	Strongly agree
a. I have received the training needed to carry out my job duties effectively.	○	○	○	○	○
b. I have the resources or equipment I need to carry out my job duties effectively.	○	○	○	○	○
c. I have the disciplinary tools needed to carry out my job duties effectively.	○	○	○	○	○
d. There are not enough MTIs to carry out all MTI duties effectively.	○	○	○	○	○

********** PAGE BREAK **********

130

Treatment of rookie MTIs (adapted from Donovan, Drasgow, and Munson, 1998): Average items a–d to form scale.

2.2 The following questions ask you about how seasoned MTIs treat newer, less-experienced rookie MTIs at BMT. Please respond based on your experiences over the last six months.

	Strongly disagree	Disagree	Neither disagree nor agree	Agree	Strongly agree
a. Seasoned MTIs help out rookie MTIs.	○	○	○	○	○
b. Seasoned MTIs argue with rookie MTIs. [Reverse scored item]	○	○	○	○	○
c. Seasoned MTIs put down rookie MTIs. [Reverse scored item]	○	○	○	○	○
d. Seasoned MTIs treat rookie MTIs with respect.	○	○	○	○	○

General MTI interpersonal treatment (adapted from Donovan, Drasgow, and Munson, 1998): Average items a–d to form scale.

2.3 The previous questions asked you about seasoned MTIs, specifically. These next questions ask about the interactions among MTIs at BMT in general. Please respond based on your experiences over the last six months.

	Strongly disagree	Disagree	Neither disagree nor agree	Agree	Strongly agree
a. MTIs help each other out.	○	○	○	○	○
b. MTIs argue with each other. [Reverse scored item]	○	○	○	○	○
c. MTIs put each other down. [Reverse scored item]	○	○	○	○	○
d. MTIs treat each other with respect.	○	○	○	○	○

********** PAGE BREAK **********

2.4 The following questions ask you about your experiences with trainees as an MTI at BMT. Please respond based on your experiences over the last six months.

At BMT. . .

	Strongly disagree	Disagree	Neither disagree nor agree	Agree	Strongly agree
a. Leaders take the word of trainees over the word of MTIs.	○	○	○	○	○
b. Leaders hold trainees accountable for making false reports against MTIs.	○	○	○	○	○
c. Trainees are honest about how MTIs treat them.	○	○	○	○	○
d. Trainees respect MTI authority.	○	○	○	○	○
e. Trainees have more power than MTIs at BMT.	○	○	○	○	○
f. I am satisfied with the overall quality of trainees.	○	○	○	○	○

********** PAGE BREAK **********

2.5 The following questions ask you about your perceptions of promotion opportunities in the Air Force as an MTI. Please read each statement carefully and indicate the extent to which you agree or disagree with the statement.

	Strongly disagree	Disagree	Neither disagree nor agree	Agree	Strongly agree
a. Those who do well as an MTI are given fair consideration for a good follow-on assignment.	○	○	○	○	○
b. MTIs are promoted to the next higher rank more slowly than NCOs serving in other assignments.	○	○	○	○	○

********** PAGE BREAK **********

Section III

3.1 The following questions ask you to think about your interactions with your immediate supervisor at BMT. Please respond based on your experiences in the last six months.

At BMT, my immediate supervisor . . .

	Strongly disagree	Disagree	Neither disagree nor agree	Agree	Strongly agree
a. Praises MTIs for doing a good job.	○	○	○	○	○
b. Plays favorites. [Reverse scored item]	○	○	○	○	○
c. Trusts MTIs.	○	○	○	○	○
d. Deals with MTIs' concerns effectively.	○	○	○	○	○
e. Treats MTIs like children. [Reverse scored item]	○	○	○	○	○
f. Treats MTIs with respect.	○	○	○	○	○
g. Responds to MTIs' questions and problems quickly.	○	○	○	○	○
h. Ignores feedback from MTIs. [Reverse scored item]	○	○	○	○	○
i. Appreciates MTIs' hard work.	○	○	○	○	○
j. Treats MTIs fairly.	○	○	○	○	○

********** PAGE BREAK **********

Ethical leadership (adapted from Brown, Trevino, and Harrison, 2005): Average items a–e to form scale.

3.2 The following questions ask you to think about your immediate supervisor at BMT and how he/she conducts him- or herself as a leader. Please respond based on your experiences in the last six months.

At BMT, my immediate supervisor . . .

	Strongly disagree	Disagree	Neither disagree nor agree	Agree	Strongly agree
a. Conducts his/her personal life in an ethical manner.	O	O	O	O	O
b. Defines success not just by results but also the way that they are obtained.	O	O	O	O	O
c. Discusses ethics or Air Force core values with MTIs.	O	O	O	O	O
d. Sets an example of how to do things the right way in terms of ethics.	O	O	O	O	O
e. Asks "what is the right thing to do?" when making decisions.	O	O	O	O	O

********** PAGE BREAK **********

Work-family conflict (Carlson, Kacmar, and Williams, 2000): Average items a–c to form work interference with family dimension; average items d–f to form family interference with work dimension.

4.1 The following questions ask you about you the extent to which your work and family lives impact each other. Family could include a spouse or other partner, children, parents, grandparents, or other extended family that play a role in your day-to-day life. Please respond based on your workload over the last six months. Please choose "not applicable" if you feel the item is not relevant to your current personal life.

	Strongly disagree	Disagree	Neither disagree nor agree	Agree	Strongly agree	Not applicable
a. When I get home from work, I am often too worn out to participate in family activities or responsibilities.	○	○	○	○	○	○
b. I am often so emotionally drained when I get home from work that it prevents me from contributing to my family.	○	○	○	○	○	○
c. Due to all the pressures at work, often when I come home I am too stressed to do the things I enjoy.	○	○	○	○	○	○
d. Due to stress at home, I am often preoccupied with family matters at work.	○	○	○	○	○	○
e. Because I am often stressed from family responsibilities, I have a hard time concentrating on my work.	○	○	○	○	○	○
f. Tension and anxiety from my family life often weaken my ability to do my job.	○	○	○	○	○	○

********** PAGE BREAK **********

4.2 The following list presents different types of stressors that people may experience as part of their job.

Please indicate the extent to which the following have caused you stress over the past six months.

	No stress	A little stress	A moderate amount of stress	Quite a bit of stress	A great deal of stress
a. Long work hours	○	○	○	○	○
b. Too little time with my family	○	○	○	○	○
c. Length of duty tour	○	○	○	○	○
d. Feeling like I am supposed to be on duty 24/7	○	○	○	○	○
e. Constant changes in policy	○	○	○	○	○
f. Unrealistic job expectations	○	○	○	○	○
g. Micromanaging of my work	○	○	○	○	○
h. Lack of opportunity to talk with leadership	○	○	○	○	○
i. Feeling like I always have to prove myself	○	○	○	○	○
j. Insufficient feedback about my performance	○	○	○	○	○
********** PAGE BREAK **********					
k. The feeling that different rules apply to different MTIs (for example, blue ropes, instructor supervisors)	○	○	○	○	○
l. Conflicting job expectations	○	○	○	○	○
m. Trainee use of comment forms for minor complaints	○	○	○	○	○
n. Working with unprofessional or incompetent MTIs	○	○	○	○	○
o. MTIs misconduct	○	○	○	○	○

136

	No stress	A little stress	A moderate amount of stress	Quite a bit of stress	A great deal of stress
p. Unfair treatment based on race, ethnicity, religion, gender, or sexual orientation.	○	○	○	○	○
q. MTIs bullying or hazing other MTIs	○	○	○	○	○
r. Inconsistent policy guidance	○	○	○	○	○

********** PAGE BREAK **********

	No stress	A little stress	A moderate amount of stress	Quite a bit of stress	A great deal of stress
s. Fear of accidently breaking a training policy	○	○	○	○	○
t. MTI lack of authority over trainees	○	○	○	○	○
u. Involuntary MTI duty extensions	○	○	○	○	○
v. Shift work	○	○	○	○	○
w. Lack of time off	○	○	○	○	○
x. Physical demands of the job	○	○	○	○	○
y. Lack of predictability in work schedule	○	○	○	○	○
z. Leadership overreactions to MTI mistakes	○	○	○	○	○
aa. Procedures for investigating accusations of MTI wrongdoing	○	○	○	○	○
bb. Competition among MTIs	○	○	○	○	○

********** PAGE BREAK **********

4.3 On average, how many hours do you work a day? [INSERT DROP DOWN BOX WITH NUMBERS 1–24.] (Note: Alternative groupings based on test survey include 7–8 hours, 9–10 hours, 11–12 hours, and 13 or more hours.)

4.4 During the past month, how many hours of actual sleep did you get in a 24-hour period? (This may be different from the number of hours you spend in bed.)

HOURS OF SLEEP PER 24-HOURS [INSERT DROP DOWN BOX WITH NUMBERS 1–24] (Note: Alternative groupings based on test survey include 5 or fewer hours, 6–7 hours, and 8 or more hours.)

4.5 During the past month, how would you rate your sleep quality overall?
- ○ Very good
- ○ Fairly good
- ○ Neither good nor bad
- ○ Fairly bad
- ○ Very bad

********** PAGE BREAK **********

137

Section V

5.1 To what extent do you believe your instructional skills have improved over the past six months?

- ○ To a great extent
- ○ To a considerable extent
- ○ To some extent
- ○ To a limited extent
- ○ To no extent

5.2 Have you taken a deliberate development course since becoming an MTI?

- ○ Yes, and I benefited from taking it.
- ○ Yes, but it wasn't very helpful.
- ○ No, but I would like to.
- ○ No, and I'm not interested.

********** PAGE BREAK **********

Bullying: (1) create a dichotomous variable indicating how many MTIs were aware of at least one item occurring versus not aware of any items occurring; (2) create an ordinal variable representing the greatest frequency that the MTI is aware of any item occurring (not that I'm aware of = 1 to 5 = daily); (3) create a dichotomous variable for awareness of each item.

6.1 Below is a list of things trainees may have done to other trainees while you were serving as an MTI at BMT. Please think about whether you are <u>personally aware</u> of any of the following things happening at BMT. Please do NOT include anything you may have only learned about from your leadership or heard about on the news.

As a reminder, for this section and all other sections that ask you about the last six months, if you have been an MTI for less time, please just answer based on the time you have been here.

<u>In the past six months, has a trainee . . .</u>

	Not that I'm aware of	Once or twice	A few times	Weekly	Daily
a. Encouraged trainees to turn against another trainee?	○	○	○	○	○
b. Tried to embarrass another trainee?	○	○	○	○	○
c. Tried to get another trainee into trouble with an MTI?	○	○	○	○	○
d. Stolen something from another trainee?	○	○	○	○	○
e. Please select "Once or twice" for this item to help us confirm that MTIs are reading these items. [Screening item]	○	○	○	○	○
f. Threatened another trainee?	○	○	○	○	○
g. Hit or kicked another trainee?	○	○	○	○	○

********** PAGE BREAK **********

Climate for Bullying (Squadron Leader Actions): Average items a–d to form scale.

6.2 BMT Trainee Rules of Conduct state that trainees are required to act in a respectful, professional manner at all times, including when interacting with other trainees. This means that bullying behaviors are not acceptable. Examples of bullying include calling another trainee insulting names, hitting another trainee, and spreading lies about a trainee.

The following questions ask you about the extent to which these rules against <u>bullying behaviors</u> are enforced at BMT. **For questions about squadron leaders, we are referring to those Air Force NCOs and officers with squadron-wide leadership responsibilities: the squadron commander, director of operations, superintendent, and first sergeant.** Please respond based on what you believe about your squadron leaders, even if you do not have direct knowledge about their attitudes or actions on this specific type of behavior.

	Strongly disagree	Disagree	Neither disagree nor agree	Agree	Strongly agree
a. Squadron leaders make honest efforts to stop bullying.	○	○	○	○	○
b. Squadron leaders encourage the reporting of bullying.	○	○	○	○	○
c. Squadron leaders take actions to prevent bullying.	○	○	○	○	○
d. Squadron leaders would correct or discipline a trainee who bullies another trainee.	○	○	○	○	○

********** PAGE BREAK **********

Maltreatment/maltraining: (1) Create a dichotomous variable indicating how many MTIs were aware of at least one item occurring versus not aware of any items occurring; (2) create an ordinal variable representing the greatest frequency that the MTI is aware of any item occurring (not that I'm aware of = 1 to 5 = daily); (3) create a dichotomous variable for awareness of any item in the following groupings:

- maltraining (a, b, c, d, f)
- privacy violations (g, h)
- denial of services or rights (i, j)
- hostile comments (l, m)
- encourage trainee mistreatment (e)
- physical threats or force (n, o, p, q, r).

7.1 Below is a list of things MTIs may have done while you were serving as an MTI at BMT. Please think about whether you are personally aware of any of the following things happening at BMT. Please do NOT include anything you may have only learned about from your leadership or heard about on the news.

In the past six months, did an MTI . . .

	Not that I'm aware of	Once or twice	A few times	Weekly	Daily
a. Discipline only one trainee when others made the same mistakes?	○	○	○	○	○
b. Unfairly push a trainee to quit or leave BMT?	○	○	○	○	○
c. Assign a trainee activities unrelated to training objectives (for example, asked a trainee to do their personal errands)?	○	○	○	○	○
d. Assign a trainee training tasks that were against the rules (for example, required PT in the latrine)?	○	○	○	○	○
e. Encourage a trainee to mistreat another trainee?	○	○	○	○	○
f. Make a trainee do PT, drill, or outside work details in unsafe conditions (for example, during black flag conditions/temperatures above 90 degrees, extreme cold weather)?	○	○	○	○	○

********** PAGE BREAK **********

	Not that I'm aware of	Once or twice	A few times	Weekly	Daily
g. Search a trainee's private mail or property for personal reasons?	○	○	○	○	○
h. Take a picture or videotape of a trainee for personal reasons?	○	○	○	○	○
i. Deny a trainee access to BMT services (for example, medical, SARC, chaplain)?	○	○	○	○	○
j. Deny a trainee other rights provided by BMT (for example, withheld their mail or refused their authorized phone call)?	○	○	○	○	○
k. Please select "A few times" for this item to help us confirm that MTIs are reading these items. [Screening item]	○	○	○	○	○
l. Call a trainee insulting names (for example, "fatso," "ugly," or "idiot")?	○	○	○	○	○
********** PAGE BREAK **********					
m. Make negative comments about a trainee's race, ethnicity, religion, gender, or sexual orientation?	○	○	○	○	○
n. Threaten to hurt a trainee?	○	○	○	○	○
o. Intentionally damage something belonging to a trainee?	○	○	○	○	○
p. Hit or punch an object when angry (for example, a wall, window, table, or other object)?	○	○	○	○	○
q. Intentionally throw something at a trainee?	○	○	○	○	○
r. Use physical force with a trainee (for example, poked, hit, grabbed, or shoved a trainee)?	○	○	○	○	○

********** PAGE BREAK **********

Climate for maltreatment/maltraining (squadron leader actions): Average items a–d to form scale.

7.2 BMT policies establish approved training methods and appropriate interactions between MTIs and trainees. MTIs making trainees perform humiliating tasks or physical exercise in unsafe conditions, threatening or hitting trainees, and using crude or offensive language are examples of policy violations that BMT calls maltreatment or maltraining.

The following questions ask you about the extent to which policies against maltreatment and maltraining are enforced at BMT. **For questions about squadron leaders, we are referring to those Air Force NCOs and officers with squadron-wide leadership responsibilities: the squadron commander, director of operations, superintendent, and first sergeant.** Please respond based on what you believe about your squadron leaders, even if you do not have direct knowledge about their attitudes or actions on this specific type of behavior.

	Strongly disagree	Disagree	Neither disagree nor agree	Agree	Strongly agree
a. Squadron leaders make honest efforts to stop maltreatment and maltraining.	○	○	○	○	○
b. Squadron leaders encourage the reporting of maltreatment and maltraining.	○	○	○	○	○
c. Squadron leaders take actions to prevent maltreatment and maltraining.	○	○	○	○	○
d. Squadron leaders would correct or discipline an MTI who engages in maltreatment or maltraining.	○	○	○	○	○

********** PAGE BREAK **********

7.3 The following questions ask you about MTIs reporting <u>maltreatment and maltraining</u> at BMT. For each statement, please <u>think about MTI behavior in general at BMT</u> and indicate the extent to which you agree or disagree with the statement.

	Strongly disagree	Disagree	Neither disagree nor agree	Agree	Strongly agree
a. MTIs would report another MTI for maltreatment or maltraining.	○	○	○	○	○
b. An MTI who reported another MTI for maltreatment or maltraining would experience retaliation from other MTIs.	○	○	○	○	○
c. MTIs would resolve most incidents of maltreatment or maltraining without making a formal report.	○	○	○	○	○
d. MTIs would help another MTI cover up an incident of maltreatment or maltraining.	○	○	○	○	○

********** PAGE BREAK **********

Section VIII

8.1 Below is a list of things <u>MTIs</u> may have done while you were serving as an MTI at BMT. Please think about whether you are <u>personally aware</u> of any of the following things happening at BMT. Please do NOT include anything you may have only learned about from your leadership or heard about on the news.

<u>In the past six months, did an MTI...</u>

		Not that I'm aware of	Once or twice	A few times	Weekly	Daily
a.	Ask a trainee to "just call me by my first name"?	o	o	o	o	o
b.	Drink alcohol with a trainee?	o	o	o	o	o
c.	Flirt with a trainee?	o	o	o	o	o
d.	Give a trainee more privileges than others even though the trainee didn't earn them?	o	o	o	o	o
e.	Contact a trainee through non-Air Force channels for personal reasons (for example, by note, phone, email, Internet, or text)?	o	o	o	o	o
f.	Share sexual jokes with a trainee?	o	o	o	o	o
g.	Meet a trainee alone?	o	o	o	o	o
h.	Talk about <u>his or her</u> sex life with a trainee?	o	o	o	o	o
i.	Talk about <u>a trainee's</u> sex life with the trainee?	o	o	o	o	o
j.	Talk about dating a trainee after the trainee graduates?	o	o	o	o	o

********** PAGE BREAK **********

	Not that I'm aware of	Once or twice	A few times	Weekly	Daily
k. Invite a trainee to a social gathering (for example, parties, or cookouts)?	o	o	o	o	o
l. Offer to give or loan a trainee money or pay for something for a trainee?	o	o	o	o	o
m. Ask a trainee to give or loan them money or buy something?	o	o	o	o	o
n. Please select "Daily" for this item to help us confirm that MTIs are reading these items. [Screening item]	o	o	o	o	o
o. Use a trainee's cell phone or other personal property?	o	o	o	o	o
p. Have a romantic relationship with a trainee?	o	o	o	o	o
q. Engage in any type of sexual activity with a trainee?	o	o	o	o	o

********** PAGE BREAK **********

Climate for unprofessional relationships: Average items a–d to form scale.

8.2 BMT policy states that MTIs are not allowed to develop friendships or romantic relationships with trainees or show favoritism to specific trainees. The Air Force deems these unprofessional relationships, even if they develop only through cards, letters, emails, phone calls, the Internet, or instant messaging. Examples of behaviors that violate Air Force professional relationship policies include an MTI giving a trainee special privileges as well as MTIs dating, drinking alcohol with, or sharing sexual stories with trainees.

The following questions ask you about the extent to which these policies against <u>unprofessional relationships</u> are enforced at BMT. **For questions about squadron leaders, we are referring to those Air Force NCOs and officers with squadron-wide leadership responsibilities: the squadron commander, director of operations, superintendent, and first sergeant.** Please respond based on what you believe about your squadron leaders, even if you do not have direct knowledge about their attitudes or actions on this specific type of behavior.

	Strongly disagree	Disagree	Neither disagree nor agree	Agree	Strongly agree
a. Squadron leaders make honest efforts to stop unprofessional relationships.	○	○	○	○	○
b. Squadron leaders encourage the reporting of unprofessional relationships.	○	○	○	○	○
c. Squadron leaders take actions to prevent unprofessional relationships.	○	○	○	○	○
d. Squadron leaders would correct or discipline an MTI who engages in an unprofessional relationship.	○	○	○	○	○

********** PAGE BREAK **********

8.3 The following questions ask you about MTIs reporting <u>unprofessional relationships</u> at BMT. For each statement, please think about MTI behavior in general at BMT and indicate the extent to which you agree or disagree with the statement.

	Strongly disagree	Disagree	Neither disagree nor agree	Agree	Strongly agree
a. MTIs would report another MTI for an unprofessional relationship with a trainee.	○	○	○	○	○
b. An MTI who reported another MTI for an unprofessional relationship with a trainee would experience retaliation from other MTIs.	○	○	○	○	○
c. MTIs would resolve most incidents of unprofessional relationships with a trainee without making a formal report.	○	○	○	○	○
d. MTIs would help another MTI cover up an unprofessional relationship with a trainee.	○	○	○	○	○

********** PAGE BREAK **********

Section IX

Sexual harassment: (1) Create a dichotomous variable indicating how many MTIs were aware of at least one item occurring versus not aware of any items occurring; (2) create an ordinal variable representing the greatest frequency that the MTI is aware of any item occurring (not that I'm aware of = 1 to 5 = daily); (3) create a dichotomous variable for awareness of any item in the following groupings:

- sexist hostility (a, b, c)
- challenges to masculinity/femininity (d, e)
- sexual hostility (f, g, h, i)
- sexual coercion (j, l, m, n)
- unwanted sexual attention (o, p, q).

9.1 Please think about whether you are <u>personally aware</u> of any of the following things happening to a trainee while you were serving as an MTI at BMT. <u>The person who did these things could be an MTI, another trainee, or someone else at BMT.</u> Focus only on things that <u>the trainee</u> found unwanted, uninvited, and in which they did not participate willingly.

Please do NOT include anything you may have only learned about from your leadership or heard about on the news.
In the past six months, has anyone, whether male or female...

		Not that I'm aware of	Once or twice	A few times	Weekly	Daily
a.	Treated a trainee "differently" because of their gender (for example, mistreated, slighted, or ignored the trainee)?	○	○	○	○	○
b.	Displayed, used, or distributed sexist or suggestive materials to a trainee (for example, pictures, stories, or pornography that the trainee found offensive)?	○	○	○	○	○
c.	Made offensive sexist remarks toward a trainee (for example, suggesting that people of the trainee's gender are not suited for the kind of work the trainee does)?	○	○	○	○	○
d.	Called a trainee gay as an insult (for example, "fag," "queer," or "dyke")?	○	○	○	○	○
e.	Insulted a trainee by saying they were not acting like a real man or real woman (for example, called the trainee a "sissy" or said the trainee was "acting like a girl" or "pretending to be a man")?	○	○	○	○	○
f.	Repeatedly told sexual stories or jokes that were offensive to a trainee?	○	○	○	○	○
g.	Made unwelcome attempts to draw a trainee into a discussion of sexual matters (for example, attempted to discuss or comment on the trainee's sex life)?	○	○	○	○	○
h.	Made gestures or used body language of a sexual nature which embarrassed or offended a trainee?	○	○	○	○	○
i.	Made offensive remarks about a trainee's appearance, body, or sexual activities?	○	○	○	○	○

********** PAGE BREAK **********

	Not that I'm aware of	Once or twice	A few times	Weekly	Daily
j. Made a trainee feel like they were being bribed with some sort of reward or special treatment to engage in sexual behavior?	○	○	○	○	○
k. Please select "Weekly" for this item to help us confirm that MTIs are reading these items. [Screening item]	○	○	○	○	○
l. Made a trainee feel threatened with some sort of retaliation for not being sexually cooperative (for example, by mentioning an upcoming test)?	○	○	○	○	○
m. Treated a trainee badly for refusing to have sex with him or her?	○	○	○	○	○
n. Implied a trainee would receive better performance evaluations or better treatment if they were sexually cooperative?	○	○	○	○	○
o. Made unwanted attempts to establish a sexual relationship with a trainee despite their efforts to discourage it?	○	○	○	○	○
p. Touched a trainee in a way that made the trainee uncomfortable?	○	○	○	○	○
q. Made unwanted attempts to touch or kiss a trainee?	○	○	○	○	○

********** PAGE BREAK **********

Climate for sexual harassment (squadron leader actions): Average items a–d to form scale.

9.2 Air Force policy states: "Unwelcome sexual advances, requests for sexual favors, and other verbal or physical conduct of a sexual nature constitute sexual harassment when (1) submission to such conduct is made either explicitly or implicitly a term or condition of an individual's employment, (2) submission to or rejection of such conduct by an individual is used as the basis for employment decisions affecting such individual, or (3) such conduct has the purpose or effect of unreasonably interfering with an individual's work performance or creating an intimidating, hostile, or offensive working environment."

The following questions ask you about the extent to which <u>sexual harassment</u> policies are enforced at BMT. **For questions about squadron leaders, we are referring to those Air Force NCOs and officers with squadron-wide leadership responsibilities: the squadron commander, director of operations, superintendent, and first sergeant.** Please respond based on what you believe about your squadron leaders, even if you do not have direct knowledge about their attitudes or actions on this specific type of behavior.

	Strongly disagree	Disagree	Neither disagree nor agree	Agree	Strongly agree
a. Squadron leaders make honest efforts to stop sexual harassment.	○	○	○	○	○
b. Squadron leaders encourage the reporting of sexual harassment.	○	○	○	○	○
c. Squadron leaders take actions to prevent sexual harassment.	○	○	○	○	○
d. Squadron leaders would correct or discipline someone who engages in sexual harassment.	○	○	○	○	○

********** PAGE BREAK **********

9.3 The following questions ask you about MTI reporting of <u>sexual harassment</u> at BMT. For each statement, please think about <u>MTI behavior in general</u> at BMT and indicate the extent to which you agree or disagree with the statement.

	Strongly disagree	Disagree	Neither disagree nor agree	Agree	Strongly agree
a. MTIs would report another MTI for sexually harassing a trainee.	○	○	○	○	○
b. An MTI who reported another MTI for sexually harassing a trainee would experience retaliation from other MTIs.	○	○	○	○	○
c. MTIs would resolve most incidents of trainees being sexually harassed without making a formal report.	○	○	○	○	○
d. MTIs would help another MTI cover up an incident of a trainee being sexually harassed.	○	○	○	○	○

********** PAGE BREAK **********

152

Section X

Unwanted sexual experiences: Analyze items a–b and d–e separately.

10.1 The following questions ask about unwanted sexual experiences that might have happened to trainees while they were at BMT. The person who did these things could be an MTI, another trainee, or someone else at BMT.

In the past six months, were you personally aware of any of the following situations?

Please do NOT include anything you may have only learned about from your leadership or heard about on the news.

	No, I'm not personally aware of this happening	Yes, I am personally aware of this happening
a. An unwanted sexual experience in which someone showed a trainee the private areas of their body, or made a trainee show them private areas of the trainee's body? (By *private areas* we mean vagina or penis, anus, groin, breast, inner thigh, and buttocks.)	○	○
b. An unwanted sexual experience in which someone touched, kissed, or rubbed up against a trainee's private area?	○	○
c. Please select "Yes, I am personally aware of this happening" for this item to help us confirm that MTIs are reading these items. **[Screening item]**	○	○
d. An unwanted sexual experience in which someone had oral, vaginal, or anal sex with a trainee? (Penetration of the vagina or anus by a penis, fingers, or any object is considered sex. Oral sex is anytime someone puts their mouth on someone's vagina or penis [even if ejaculation does not occur]).	○	○
e. An unwanted sexual experience in which someone TRIED but failed to have oral, vaginal, or anal sex with a trainee?	○	○

********** PAGE BREAK **********

10.2 The Uniform Code of Military Justice (UCMJ) criminalizes various forms of unwanted sexual activity, including rape, sexual assault, and unwanted sexual contact. For the next sections, we use the term *sexual assault* to refer to all of these forms of unwanted sexual contact characterized by force, threats, intimidation, or abuse of authority, or when the victim does not or cannot consent to that sexual contact.

The following questions ask you about the extent to which military laws and policies on sexual assault are enforced at BMT. **For questions about squadron leaders, we are referring to those Air Force NCOs and officers with squadron-wide leadership responsibilities: the squadron commander, director of operations, superintendent, and first sergeant.** Please respond based on what you believe about your squadron leaders, even if you do not have direct knowledge about their attitudes or actions on this specific type of behavior.

	Strongly disagree	Disagree	Neither disagree nor agree	Agree	Strongly agree
a. Squadron leaders make honest efforts to stop sexual assault.	○	○	○	○	○
b. Squadron leaders encourage the reporting of sexual assault.	○	○	○	○	○
c. Squadron leaders take actions to prevent sexual assault.	○	○	○	○	○
d. Squadron leaders would discipline someone who engages in sexual assault.	○	○	○	○	○

********** PAGE BREAK **********

10.3 The following questions ask you about MTIs reporting <u>sexual assault</u> at BMT. For each statement, please think about MTI behavior in general at BMT and indicate the extent to which you agree or disagree with the statement.

	Strongly disagree	Disagree	Neither disagree nor agree	Agree	Strongly agree
a. MTIs would report another MTI for sexually assaulting a trainee.	○	○	○	○	○
b. An MTI who reported another MTI for sexually assaulting a trainee would experience retaliation from other MTIs.	○	○	○	○	○
c. MTIs would resolve most incidents of trainees being sexually assaulted without making a formal report.	○	○	○	○	○
d. MTIs would help another MTI cover up the sexual assault of a trainee.	○	○	○	○	○

********** PAGE BREAK **********

Section XI

11.1 The following questions ask you about the clarity of policies at BMT. Please read each statement carefully and indicate the extent to which you agree or disagree with the statement.

	Strongly disagree	Disagree	Neither disagree nor agree	Agree	Strongly agree
a. BMT trainee rules of conduct are clear.	○	○	○	○	○
b. BMT maltreatment and maltraining policies are clear.	○	○	○	○	○
c. AETC unprofessional relationship policies are clear.	○	○	○	○	○
d. Air Force sexual harassment policies are clear.	○	○	○	○	○
e. Air Force sexual assault laws and policies are clear.	○	○	○	○	○

********** PAGE BREAK **********

Closing Questions

How open and honest did you feel you could be when answering these survey questions?

- ○ Not at all open and honest
- ○ Somewhat open and honest
- ○ Completely open and honest

Would you like to tell us anything more that would clarify how you feel about your working conditions, your leadership, or quality of life as an MTI? [INSERT OPEN-ENDED TEXT BOX]

********** PAGE BREAK **********

THANK YOU FOR PARTICIPATING IN THIS SURVEY.

********** PAGE BREAK **********

Appendix D. Sample Page from Reporting Template for Trainee Survey

(See table on next page.)

Selected yes to any in Q1.1: Experienced Any Bullying			Q1.1 Individual Bullying Items Ever Experienced						Greatest frequency with which any bullying item was experienced				
Total in the analytic sample (excludes missing)			Q1.1a Turn Against	Q1.1b Embarrass	Q1.1c Trouble w/MTI	Q1.1d Steal	Q1.1f Threaten	Q1.1g Hit/ Kick	Never (No items in Q1.1 experienced)	Any item experienced "Once or twice" but none more frequently	Any item experienced "A few times" but none more frequently	Any item experienced "Weekly" but none "Daily"	Any item in Q1.1 experienced "Daily"
Yes		Total N											
N	%		N	N	N	N	N	N	N	N	N	N	N
FY14_1													
FY14_2													
FY14_3													
FY14_4													

Appendix E. Proposed Core Content for an MTI Survey Recruitment Letter

This appendix contains our proposed invitation to be sent to all MTIs at BMT. We suggest sending it even to MTIs not requested to participate (those who have been at BMT less than one month) so they are aware of what is happening and to promote open communication.

The letter could be tailored with additional content. AETC might also want to mention changes that resulted from previous MTI surveys, so that MTIs can see that leadership takes these surveys seriously and acts to address MTI concerns. It might also mention the particular context, such as:

- We recognize that we have been shorthanded the past six months, and this survey is one way we'll be assessing any potential impacts on MTIs
- We've made a number of changes at BMT over the past year, and MTI feedback through this survey is one way we'll be assessing what's been successful and what we still need to address.
- A lot of new MTIs have arrived at BMT since our last MTI survey, so we're looking forward to hearing from many of you for the first time, as well as continuing to gather input from our more seasoned MTIs.

As consistent core content, we recommend the following statement.

AETC MTI Survey Recruitment Statement

All MTIs who have served as an MTI for at least one month are asked to participate in an anonymous survey designed to assess MTI work experiences, quality of life, and the BMT environment. Your responses from this survey will be combined with other MTIs' responses and shared with BMT leaders so they can better understand the issues MTIs face in the training environment and where there might be problems that leaders need to address.

The survey will be completely anonymous. It will not ask for your name, an identification number, or your contact information. Your CAC will not be needed to take the survey. Also, your participation in this survey is completely voluntary. This means you may skip questions or choose to stop taking the survey at any time.

There is no penalty if you choose not to participate. Participation will not help or harm your future assignments or promotions in the Air Force.

Your responses are important in helping to provide feedback to Air Force leaders about MTI working conditions as well as abuse or misconduct at BMT so they can take additional steps to ensure a positive and safe training environment.

Although the survey sessions have been scheduled in one-hour time blocks, the survey should take less time to complete. Participants are asked to please arrive at the beginning of the time slot to receive the survey introduction.

The survey will be held in the [classroom facility, room number]. MTIs can choose to attend any one of the following time slots on [month and day] or [alternative month and day]:

- 7:30
- 10:30
- 12:30
- 14:00
- 15:30

Thank you in advance for your participation.

Appendix F. Developing an Integrated Feedback System for Addressing Abuse and Misconduct

This appendix describes other data sources and feedback mechanisms at BMT that RAND reviewed to determine how the surveys could fit into a broader integrated feedback system that would be more effective in combatting abuse and misconduct. In the sections that follow, we describe these additional data sources and how each source can provide feedback on abuse and misconduct. We specifically focus on eight domains of data presented according to whom the data are about and from whom the data are collected: trainee data, MTI data, general population surveys, official incident data, hotline data, chaplain data, SARC data, BMT production data, and security camera surveillance. Subsequently, we discuss how the data, along with the newly developed surveys described in this report, can be integrated into a more comprehensive feedback.

Trainee Data

Trainee data includes information about trainees' performance, the training environment, and indicators of health and well-being that can be observed at the individual-trainee level, and then extrapolated to the flight or squadron level. It may be reported by the trainee or observed by others. Some information can be attributed to an individual, while other sources are anonymous. We have identified these existing sources of trainee data:

- **BMT Mental Health Screening** includes the Lackland Behavioral Questionnaire and a measure of resilience. It is completed by trainees upon entering basic training, usually within 72 hours of arrival.
- The **End-of-Course Survey** contains survey items rating various aspects of the BMT experience, including whether instructors meet core competencies, whether policies are complied with, general perceptions of the training experience, and satisfaction with services (e.g., cafeteria, laundry, mail). It also now contains direct questions about experiences with sexual misconduct and assault, and it is not anonymous.
- **Trainee comment sheets** contain items for trainees to rate (positive or negative) aspects of their training (e.g., clothing issue, sexual harassment) and make suggestions. These sheets are placed in critique drop boxes in the cafeteria, dormitories, and elsewhere on the training campus so trainees can fill one out anonymously and drop it in the box. These sheets are collected from the drop boxes three times a week, categorized, scanned, emailed, and loaded into a tracking database.
- The BMT program evaluations analyst created and administers the **Trainee Safety and Well-Being Survey**, which is given opportunistically or strategically to various flights of trainees. It contains items that allow trainees to identify abuse and misconduct, policy

163

violations, and unprofessional relationships that they experienced, witnessed, or heard about.

Taken together, these data sources can provide useful information for addressing abuse and misconduct. However, none of the data sources provides a systemic and confidential assessment of abuse and misconduct, data on barriers preventing trainees from reporting abuse, or information about individuals' experiences following a report. Given that many trainees do not report an incident, data from official reports are likely to underestimate the true prevalence of abuse and misconduct. To help leaders understand what they can do to address underreporting, it is critical to collect data on potential barriers to reporting. The RAND survey is designed to fill these gaps and provide more-accurate prevalence estimates and feedback on trainee experiences through a systematic and confidential survey given to all trainees at BMT.

MTI Data

MTI data includes any information about instructor performance, training, and health and well-being that can be observed at the individual-MTI level. It may be reported by the instructor or observed by others. The information need not be attributable—that is, some information about MTIs may be anonymous. We have identified five existing sources of MTI data:

- **MTI screening data** includes all the data required for an airman to become an MTI, including an application form, a spouse interview, recommendations, an applicant mental health screening and interview, and review of career records.
- The **MTI Quality of Life Survey** is designed to assess MTIs' satisfaction with their careers and the organizational climate. This survey focus has been integrated into the RAND survey for MTIs.
- **Personal information files** document general misconduct, substandard performance, and other derogatory information. Examples of conduct reported in a personal information file include memorandums for the record, no contact orders, letters of counseling, letters of admonition, letters of reprimand, and disciplinary and decertification actions that are specific to certain career fields.
- The **MTI End of Course Survey** is similar to the trainee version. It asks airmen who are training to be MTIs to rate various aspects of their training at the end of every course.
- **Manning data** includes the number of authorized instructor slots, the number of filled slots, and the demographics of the instructors who fill the slots (e.g., rank, years of service, time on station, gender).

The MTI data reviewed here can be used in a number of ways. For example, manning data can provide leadership with a sense of the MTI workload and the correlated levels of stress. Additionally, the End-of-Course Survey provides an avenue for understanding whether MTIs feel that they have received the training needed to be effective in their jobs and the extent to which policies for appropriate training and discipline tools have been clearly communicated.

The MTI survey developed by RAND further complements these data sources by providing an anonymous way for assessing MTIs' awareness of abuse and misconduct at BMT, their

164

perceptions of whether squadron leaders enforce related policies and laws, and the barriers MTIs face in reporting incidents. The survey also integrates many of the constructs previously included in the MTI Quality of Life Survey.

General Population Surveys

General population surveys include Air Force–wide assessments of various aspects of life as an airman. These surveys provide a sense of the general Air Force environment in which trainees are trained, except for one characteristic: Trainees are not participants. RAND identified five relevant general population surveys:

- The **Defense Equal Opportunity Climate Survey** is designed to help commanders assess their units' human relations climate and provide insight into both positive and negative factors that may affect unit effectiveness and cohesion. The survey is Air Force–wide, except for trainees (although MTIs are included).
- The **Air Force Climate Survey** is Air Force–wide. The overall results are reported to senior leadership, and commanders at all levels receive reports of their own units' climate. It also does not survey trainees.
- The **Air Force Community Assessment Survey** is designed to assess the association between aspects of the Air Force community and outcomes such as retention, readiness, satisfaction, and cohesion. Results from the survey help advise leadership on the needs of airmen and their families. The survey uses a simple stratified random sampling technique to capture airmen at all bases but does not include trainees.
- In 2010, the Air Force Sexual Assault Prevention and Response Office directed Gallup to field the **2010 Gallup Air Force Personnel Safety Survey** to help estimate the incidence and prevalence of sexual assault in the Air Force. This is the one general population survey that included trainees, but this survey might not be repeated.
- The 2012 **Quadrennial Workplace and Gender Relations Survey of Active-Duty Members** was designed to enhance understanding of sexual assault in the military and how DoD's prevention efforts may have affected the incidence and reporting of sexual assaults. This survey was an iteration of numerous efforts by DoD since 1988 to assess these topics (another version was administered in 2014). Trainees were not included because participation required at least six months of service.

These population surveys can provide useful information, but there are some limitations. For example, they are not always completed with regularity or frequency. Some are conducted annually while others occur on an ad hoc basis, often several years apart. Not all surveys are recurring. In addition, population surveys are sometimes designed to be helpful at the aggregate level—by base, for example. Because the Community Assessment Survey is focused at the installation level, installation commanders frequently use it to assess the needs of their specific airmen and families. Thus, general population surveys may have limited utility for breaking out results specifically for BMT personnel.

Despite these limitations, these general population surveys can still be useful for a feedback system on abuse and misconduct. First, they can provide information on prevalence rates of abuse and mistreatment across the Air Force (of course, not all types of BMT-specific abuse and misconduct would be assessed elsewhere). Second, they can give leadership a sense of the culture in which airmen work and train future airmen.

Official Incident Data

Official incident data can help AETC understand the behavior of instructors, especially behavior that should be flagged as an indicator of future potential abuse or misconduct. Two sources of data are described below.

- Official **law enforcement records** include reports of formal investigations by the Air Force Office of Special Investigations and Security Forces.
- The **Automated Military Justice Analysis and Management System** is a judge advocate–managed system for reporting misconduct and case management of those reports.

In addition to serving as red flags for future negative conduct, data from these systems can be used to assess the level or prevalence of reported mistreatment and maltraining. The results from the RAND BMT survey can be compared with these reported incident rates to provide leadership a sense of what incidents may not be being reported and reasons for lack of reporting.

Hotline Data

Trainees have access to various means of reporting abuse and misconduct. One of those means is through hotlines that link trainees to different sources of support. RAND identified two hotlines operating at BMT at the time of this study:[16]

- The **24/7 2AF Line** serves as both an anonymous tip line and a means to officially report allegations of abuse and misconduct. However, it is not a venue for restricted reporting, where victims can receive confidential treatment. Operators can refer callers to the base SARC, chaplains, or the DoD SAFE Helpline, which is a phone (both voice and text) and online confidential assistance program for military victims of sexual assault.[17]
- **BMT dormitory hotline phones** have been installed in all BMT trainee housing. The phones have a simple push-button process where one line connects directly to the 24/7 2AF Line, one line connects directly to the local SARC, and one line connects directly to the chaplain.

---

[16] The Abuse/Misconduct Hotline was stood up by the commander-directed investigation (CDI) in July 2012 to facilitate reporting allegations of abuse and misconduct by BMT trainees or TT students. It ceased operation in November of 2012, when the 24/7 AF Line took over. The CDI Abuse/Misconduct Hotline referred callers to the base SARC, chaplains, or the DoD SAFE Hotline.

[17] The DoD SAFE Helpline is operated by the Rape, Abuse and Incest National Network (RAINN), which is a nonprofit organization and the country's largest anti–sexual violence group. RAINN has a contractual agreement with the DoD Sexual Assault Prevention and Response Office.

These hotlines provide the opportunity for trainees to bypass their MTIs and either report abuse and misconduct to AETC or reach out for support from a SARC, chaplain, or service dedicated to victims of sexual assault, which is then captured in the records those professionals maintain.

Chaplain Data

All trainees at BMT are supposed to have access to chaplains, who may serve as counselors, sounding boards, or spiritual leaders for trainees. Through the **Air Force Chaplain Corps Activity Reporting System**, chaplains record and track data on the issues for which they provide counseling. Chaplains record their notes after each session or encounter with a counselee, which could include information shared confidentially about incidents of abuse and misconduct, including sexual assault. According to a chaplain representative at Lackland Air Force Base, aggregate statistical data can be shared with appropriate Air Force leadership upon request, but confidential aspects of individual reports are released only with permission of the counselee.

Sexual Assault Response Coordinator (SARC) Data

All trainees should also have access to a SARC—a single point of contact for coordinating and integrating sexual assault victim care services. SARCs track the number of restricted (i.e., confidential) and unrestricted reports, as well as the details (e.g., timing, location, type of assault, circumstances of assault). Each time a SARC is contacted, a file is entered into the **SARC database**, regardless of whether the report is restricted or unrestricted. For restricted reports, very limited information is recorded, and access to that information is severely restricted.

BMT Production Data

BMT production data include graduation, attrition, and medical data. Abuse is one possible reason a trainee may fail to graduate, be recycled into a subsequent cohort,[18] or have an injury that prevents him or her from training. Because it is possible to drill down on these data to the instructor level, production data may also be useful in identifying individual MTIs who are abusing their power or employing unauthorized training methods.

- **Graduation and attrition data** include graduation rates, both on time or delayed (e.g., due to a medical hold or recycle); transfers; and attrition by entry group.

---

[18] A *recycle* refers to a trainee who leaves training flight for some reason (e.g., medical, personal) and upon his or her return picks up in the training cycle where he or she left off, regardless of where the original cohort is in the training cycle. For example, a trainee who enters a two-week medical hold at week 4 in training would reenter training at week 4, rather than week 6 with his or her original cohort.

- **Medical data** include medical holds and recycles (and reasons for medical holds), as well as injury rates, during BMT.

Production data should be viewed as a supplement to the other types of data described. The BMT program evaluations analyst reported that previous investigations into higher-than-average injury rates and/or mental health referrals in particular flights revealed that they were related to MTI abuse or improper training. Thus, these production data could be one element of a more complete feedback and monitoring system.

Security Camera Surveillance

Security camera surveillance was implemented as a result of the CDI. Camera coverage is primarily focused on locations outside of dormitories, including in stairwells, foyers and hallways and under overhangs. Any individual within range of the camera is captured, so the record could include trainees, instructors, and others. At the time of this study, the surveillance camera system stored images for 30 days, although recommendations from the CDI suggested a minimum of 45 days. The planned upgrade to the system would allow it to store data for up to two years. During our review we learned that the footage was only being examined if a complaint had been registered.

If manpower permits, those recordings should be regularly monitored so that leaders could be alerted in a timely manner to suspicious behavior that might be going unreported. This would provide another means to detect incidents that might not otherwise be reported. In addition to abuse and misconduct, this footage may reveal other prohibited behavior, such as theft.

Building an Integrated Feedback System

Along with the newly developed surveys described in this report, AETC has a variety of data sources and feedback mechanisms from which it can draw to address abuse and misconduct at BMT. Although each data source provides useful information on its own, when integrated together, these data sources can provide a much more comprehensive picture of the BMT environment and the potential issues leaders need to address.

For example, several data sources provide information on reported and unreported incidents of abuse and misconduct, as well as feedback on general awareness of the levels of abuse and misconduct taking place at BMT. Data from the various reporting channels (e.g., hotlines, SARC, trainee comment sheets) can also be combined with results from the newly developed trainee survey to help gain a better understanding of what reporting avenues trainees feel most comfortable using and to help make sure trainees have necessary access. The trainee survey also provides insight into other potential barriers for reporting and the extent to which trainees who have reported an incident have had a positive experience.

Data collected from and about MTIs (e.g., manning) provide important feedback about the extent to which MTIs have the necessary training and resources to be effective. They also provide insight into other stressors and work environment issues MTIs may be facing that could influence their ability to prevent and respond to abuse and misconduct.

General population surveys as well as BMT specific surveys can provide leadership a sense of the training environment's climate and culture. This also includes the extent to which both trainees and MTIs perceive that leadership enforces policies and encourages the reporting of abuse and misconduct. A strong climate of reporting and adherence to core Air Force values and policies may be enough to deter some individuals from abuse or misconduct. Similarly, if trainees feel that they will not be taken seriously or even feel discouraged from reporting, they may be less likely to report, which affects the ability to deter and detect abuse and misconduct.

Finally, if those found guilty of abuse or misconduct are not held accountable for their behavior, then dissuasion, deterrence, and detection have no teeth. For accountability to work, the data we reviewed must be shared with the appropriate individuals and offices. If abuse and misconduct remain unknown to those who have the authority and power to punish and discipline offenders, the feedback system will fail. The feedback system, through all the various data sources, can help BMT and Air Force leadership dissuade some would-be perpetrators, deter those who are not dissuaded, detect those who ultimately do commit such acts, and hold offenders accountable for their actions.

Bibliography

Air Education and Training Command Instruction 36-2216, *Administration of Military Standards and Discipline Training*, December 6, 2010.

Air Education and Training Command Instruction 36-2909, *Professional and Unprofessional Relationships*, March 2, 2007.

Air Force Instruction 36-2909, *Professional and Unprofessional Relationships*, May 1, 1999.

Allen, N. J., and J. P. Meyer, "The Measurement and Antecedents of Affective, Continuance and Normative Commitment to the Organization," *Journal of Occupational Psychology*, Vol. 63, No. 1, March 1990, pp. 1–18.

Bennett, R. J., and S. L. Robinson, "Development of a Measure of Workplace Deviance," *Journal of Applied Psychology*, Vol. 85, No. 3, June 2000, pp. 349–360.

Black, M. C., K. C. Basile, M. J. Breiding, S. G. Smith, M. L. Walters, M. T. Merrick, J. Chen, and M. R. Stevens, *The National Intimate Partner and Sexual Violence Survey (NISVS): 2010 Summary Report*, Atlanta: National Center for Injury Prevention and Control, November 2011. As of June 6, 2012:
http://www.cdc.gov/ViolencePrevention/pdf/NISVS_Report2010-a.pdf

Bowling, N. A., "Is the Job Satisfaction–Job Performance Relationship Spurious? A Meta-Analytic Examination," *Journal of Vocational Behavior*, Vol. 71, No. 2, 2007, pp. 167–185.

Bowling, N. A., K. J. Eschleman, and Q. A. Wang, "A Meta-Analytic Examination of the Relationship Between Job Satisfaction and Subjective Well-Being," *Journal of Occupational and Organizational Psychology*, Vol. 83, No. 4, December 2010, pp. 915–934.

Brown, M. E., L. K. Trevino, and D. A. Harrison, "Ethical Leadership: A Social Learning Perspective for Construct Development and Testing," *Organizational Behavioral and Human Decision Processes*, Vol. 97, 2005, pp. 117–134.

Browne, M. W., and R. Cudeck, "Alternative Ways of Assessing Model Fit," in K. Bollen and J. Long, eds., *Testing Structural Equation Models, Sage Focus Editions*, Vol. 154, 1993, pp. 136–162.

Buss, A. H., and M. Perry, "The Aggression Questionnaire," *Journal of Personality and Social Psychology*, Vol. 63, No. 3, September 1992, pp. 452–459.

Carlson, D. S., K. M. Kacmar, and L. J. Williams, "Construction and Initial Validation of Multidimensional Measure of Work-Family Conflict," *Journal of Vocational Behavior*, Vol. 56, 2000, pp. 246–276.

Christian, M. S., J. C. Bradley, J. C. Wallace, and M. J. Burke, "Workplace Safety: A Meta-Analysis of the Roles of Person and Situation Factors," *Journal of Applied Psychology*, Vol. 94, No. 5, 2009, pp. 1103–1127.

Cohen, J., *Statistical Power Analysis for the Behavioral Sciences*, revised ed., New York: Academic Press, 1977.

Cohen-Charash, Y., and P. E. Spector, "The Role of Justice in Organizations: A Meta-Analysis," *Organizational Behavior and Human Decision Processes*, Vol. 86, No. 2, 2001, pp. 278–321.

Colquitt, J. A., D. E. Conlon, M. J. Wesson, C. O. L. H. Porter, and K. Y. Ng, "Justice at the Millennium: A Meta-Analytic Review of 25 Years of Organizational Justice Research," *Journal of Applied Psychology*, Vol. 86, No. 3, 2001, pp. 425–445.

Congressional Commission on Military Training and Gender-Related Issues, *Congressional Commission on Military Training and Gender-Related Issues: Final Report*, Vol. 3: *Research Projects, Reports, and Studies*, Washington D.C., July 1999.

Cooper-Hakim, A., and C. Viswesvaran, "The Construct of Work Commitment: Testing an Integrative Framework," *Psychological Bulletin*, Vol. 131, No. 2, March 2005, pp. 241–259.

Cortina, L. M., V. J. Magley, J. H. Williams, and R. D. Langhout, "Incivility in the Workplace: Incidence and Impact," *Journal of Occupational Health Psychology*, Vol. 6, No. 1, 2001, p. 64.

Cowie, H., P. Naylor, I. Rivers, P. K. Smith, and B. Pereira, "Measuring Workplace Bullying," *Aggression and Violent Behavior*, Vol. 7, 2002, pp. 33–51.

Crawford, C. B., and Ferguson, G. A., "A General Rotation Criterion and Its Use in Orthogonal Rotation," *Psychometrika*, Vol. 35, 1970, pp. 321–332.

Cropanzano, R., A. Li, and L. Benson III, "Peer Justice and Teamwork Process," *Group & Organization Management*, Vol. 36, 2011, pp. 567–596.

Culbertson, A., and W. Rodgers, "Improving Managerial Effectiveness in the Workplace: The Case of Sexual Harassment of Navy Women," *Journal of Applied Social Psychology*, Vol. 27, No. 22, 1997, pp. 1953–1971.

Culhane, E., *Annual Demographic Profile of the Department of Defense and United States Coast Guard Fiscal Year 2013*, Patrick Air Force Base, Fla.: Defense Equal Opportunity Management Institute, June 2014. As of November 4, 2014: http://www.deomi.org/downloadableFiles/DOD_USCG_Deomographics_FY_2013_Final_201406021.pdf

Defense Manpower Data Center, *Service Academy 2006 Gender Relations Survey*, DMDC Report No. 2006-016, December 2006.

Department of the Army Inspector General, Special Investigation of Initial Entry Training, Equal Opportunity and Sexual Harassment Policies and Procedures December 1996–April 1997, July 22, 1997.

DoD—*See* U.S. Department of Defense.

Donovan, M. A., F. Drasgow, and L. J. Munson, "The Perceptions of Fair Interpersonal Treatment Scale: Development and Validation of a Measure of Interpersonal Treatment in the Workplace," *Journal of Applied Psychology*, Vol. 83, No. 5, 1998, pp. 683–692.

Eatough, E. M., C. H. Chang, S. A. Miloslavic, and R. E. Johnson, "Relationships of Role Stressors with Organizational Citizenship Behavior: A Meta-Analysis," *Journal of Applied Psychology*, Vol. 96, No. 3, May, 2011, pp. 619–632.

Faragher, E. B., M. Cass, and C. L. Cooper, "The Relationship Between Job Satisfaction and Health: A Meta-Analysis," *Occupational and Environmental Medicine*, Vol. 62, No. 2, February 2005, pp. 105–112.

Fisher, B. S., "The Effects of Survey Question Wording on Rape Estimates Evidence from a Quasi-Experimental Design," *Violence Against Women*, Vol. 15, No. 2, 2009, pp. 133–147.

Fisher, B. S., F. T. Cullen, and M. G. Turner, *The Sexual Victimization of College Women*, Research Report, Washington, D.C.: National Institute of Justice and the Bureau of Justice Statistics, Office of Justice Programs, U.S. Department of Justice, 2000. As of December 16, 2014:
https://www.ncjrs.gov/pdffiles1/nij/182369.pdf

Fisher, B. S., L. E. Daigle, F. T. Cullen, and M. G. Turner, "Reporting Sexual Victimization to the Police and Others: Results from a National-Level Study of College Women," *Criminal Justice and Behavior*, Vol. 30, February 2003, pp. 6–38.

Fitzgerald, L. F., M. J. Gelfand, and F. Drasgow, "Measuring Sexual Harassment: Theoretical and Psychometric Advances," *Basic and Applied Social Psychology*, Vol. 17, 1995, pp. 425–445.

Fitzgerald, L. F., V. J. Magley, F. Drasgow, and C. R. Waldo, "Measuring Sexual Harassment in the Military: The Sexual Experiences Questionnaire (SEQ-DoD)," *Military Psychology*, Vol. 11, 1999, pp. 243–263.

Fox, S., P. E. Spector, and D. Miles, "Counterproductive Work Behavior (CWB) in Response to Job Stressors and Organizational Justice: Some Mediator and Moderator Tests for Autonomy and Emotions," *Journal of Vocational Behavior*, Vol. 59, No. 3, December 2001, pp. 291–309.

Gilboa, S., A. Shirom, Y. Fried, and C. Cooper, "A Meta-Analysis of Work Demand Stressors and Job Performance: Examining Main and Moderating Effects," *Personnel Psychology*, Vol. 61, No. 2, 2008, pp. 227–271.

Goffman, E., *Asylums: Essays on the Social Situation of Mental Patients and Other Inmates*, New York: Doubleday Anchor, 1961.

Greenhaus, J. H., and N. J. Beutell, "Sources of Conflict Between Work and Family Roles," *Academy of Management Review*, Vol. 10, No. 1, 1985, pp. 76–88.

Griffeth, R. W., P. W. Hom, and S. Gaertner, "A Meta-Analysis of Antecedents and Correlates Of Employee Turnover: Update, Moderator Tests, and Research Implications for the Next Millennium," *Journal of Management*, Vol. 26, No. 3, 2000, pp. 463–488.

Gruber, J. E., "A Typology of Personal and Environmental Sexual Harassment: Research and Policy Recommendations for the 1990s," *Sex Roles*, Vol. 26, 1992, pp. 447–464.

Gruys, M. L., and P. R. Sackett, "Investigating the Dimensionality of Counterproductive Work Behavior," *International Journal of Selection and Assessment*, Vol. 11, No. 1, March 2003, pp. 30–42.

Gutek, B. A., S. Searle, and L. Klepa, "Rational Versus Gender Role Explanations for Work-Family Conflict," *Journal of Applied Psychology*, Vol. 76, No. 4, 1991, pp. 560–568.

Hamburger M. E., K. C. Basile, A. M. Vivolo, *Measuring Bullying Victimization, Perpetration, and Bystander Experiences: A Compendium of Assessment Tools*, Atlanta: Centers for Disease Control and Prevention, National Center for Injury Prevention and Control, 2011.

Herbert, T. B., and C. Dunkel-Schetter, "Negative Social Reactions to Victims: An Overview of Responses and Their Determinants," in Leo Montada, Sigrun-Heide Filipp, and Melvin J. Learner, eds., *Life Crises and Experiences of Loss in Adulthood*, Hillsdale, N.J.: Lawrence-Erlbaum Associates, April 1992, pp. 497–518.

Hershcovis, M. S., N. Turner, J. Barling, K. A. Arnold, K. E. Dupre, M. Inness, M. M. LeBlanc, and N. Sivanathan, "Predicting Workplace Aggression: A Meta-Analysis," *Journal of Applied Psychology*, Vol. 92, No. 1, 2007, pp. 228–238.

Highhouse, S., and A. S. Becker, "Facet Measures and Global Job Satisfaction," *Journal of Business and Psychology*, Vol. 8, No. 1, 1993, pp. 117–127.

Hinkin, T. R., "A Brief Tutorial on the Development of Measures for Use in Survey Questionnaires," *Organizational Research Methods*, Vol. 1, 1998, pp. 104–121.

Hu, L. T., and P. Bentler, "Cutoff Criteria for Fit Indexes in Covariance Structure Analysis: Conventional Criteria Versus New Alternatives," *Structural Equation Modeling: A Multidisciplinary Journal*, Vol. 6, No. 1, 1999, pp. 1–55.

Hulin, C. L., L. F. Fitzgerald, and F. Drasgow, "Organizational Influences on Sexual Harrassment," in M.S. Stockdale, ed., *Sexual Harassment in the Workplace: Perspectives, Frontiers, and Response Strategies*, London: Sage Publications, 1996, pp. 127–150.

Jaros, S. J., J. M. Jermier, J. W. Koehler, and T. Sincich, "Effects of Continuance, Affective, and Moral Commitment on the Withdrawal Process—an Evaluation of 8 Structural Equation Models," *Academy of Management Journal*, Vol. 36, No. 5, October 1993, pp. 951–995.

Judge, T. A., C. J. Thoresen, J. E. Bono, and G. K. Patton, "The Job Satisfaction-Job Performance Relationship: A Qualitative And Quantitative Review," *Psychological Bulletin*, Vol. 127, No. 3, May 2001, pp. 376–407.

Karrasch, A. I., "Antecedents and Consequences of Organizational Commitment," *Military Psychology*, Vol. 15, No. 3, 2003, pp. 225–236.

Kessler, S. R., P. E. Spector, C. H. Chang, and A. D. Parr, "Organizational Violence Climate and Exposure to Violence and Verbal Aggression," *Work & Stress*, Vol. 22, 2008, pp. 108–124.

Kinicki, A. J., F. M. McKee-Ryan, C. A. Schriesheim, and K. P. Carson, "Assessing the Construct Validity of the Job Descriptive Index: A Review and Meta-Analysis," *Journal of Applied Psychology*, Vol. 87, No. 1, February 2002, pp. 14–32.

Kirkpatrick, D. L., and J. D. Kirkpatrick, *Evaluating Training Programs: The Four Levels*, 3rd ed., San Francisco: Berrett-Koehler Publishers, 2006.

Kolivas, E. D., and A. M. Gross, "Assessing Sexual Victimization: Addressing the Gap Between Rape Victimization and Perpetration Prevalence Rates," *Aggression and Violent Behavior*, Vol. 12, 2007, pp. 315–328.

Kopelman, R. E., J. H. Greenhaus, and T. F. Connolly, "A Model of Work, Family, and Interrole Conflict: A Construct Validation Study," *Organizational Behavior and Human Performance*, Vol. 32, 1983, pp. 198–215.

Koss, M. P., "Detecting the Scope of Rape: A Review of Prevalence Research Methods," *Journal of Interpersonal Violence*, Vol. 8, 1993, pp. 198–222.

Koss, M. P., C. A. Gidycz, and N. Wisniewski, "The Scope of Rape: Incidence and Prevalence of Sexual Aggression and Victimization in a National Sample of Higher Education Students," *Journal of Consulting and Clinical Psychology*, Vol. 55, 1987, pp. 162–170.

Koss, M. P., and C. J. Oros, "Sexual Experiences Survey: A Research Instrument Investigating Sexual Aggression and Victimization," *Journal of Consulting and Clinical Psychology*, Vol. 50, 1982, pp. 455–457.

Kruttschnitt, Candace, William D. Kalsbeek, and Carol C. House, eds., *Estimating the Incidence of Rape and Sexual Assault*, Washington, D.C.: National Academies Press, 2014.

Latour, S. M., and S. K. Marston, "Every Citizen a Soldier: Historic Foundations for Gender Integrated Training (GIT) and Implications for Air Force Readiness," Montgomery, Ala.: Air Command and Staff College, Air University, Maxwell Air Force Base, April 1999.

Lewis, S. F., and W. Fremouw, "Dating Violence: A Critical Review of the Literature," *Clinical Psychology Review*, Vol. 21, No. 1, 2001, pp. 105–127.

Li, A., and R. Cropanzano, "Fairness at the Group Level: Interunit and Intraunit Justice Climate," *Journal of Management*, Vol. 35, 2009, pp. 564–599.

Locke, E. A., "The Nature and Causes of Job Satisfaction," in M. D. Dunette, ed., *Handbook of Industrial and Organizational Psychology*, Chicago: Rand McNally, 1976, pp. 1297–1349.

Mahmood, M. H., S. J. Coons, M. C. Guy, and K. R. Pelletier, "Development and Testing of the Workplace Stressors Assessment Questionnaire," *Journal of Occupational and Environmental Medicine*, Vol. 52, No. 12, December 2010, pp. 1192–1200.

Mayer, R. C., and F. D. Schoorman, "Predicting Participation and Production Outcomes Through a Two-Dimensional Model Of Organizational Commitment," *Academy of Management Journal*, Vol. 35, No. 3, 1992, pp. 671–684.

———, "Differentiating Antecedents of Organizational Commitment: A Test of March and Simon's Model," *Journal of Organizational Behavior*, Vol. 19, No. 1, January 1998, pp. 15–28.

McWhortner, S. K., V. A. Stander, L. L. Merrill, C. J. Thomsen, and J. S. Milner, *Reports of Rape Reperpetration by Newly Enlisted Male Navy Personnel*, San Diego, Calif.: Naval Health Research Center, 2009.

Meyer, J. P., and L. Herscovitch, "Commitment in the Workplace: Toward a General Model," *Human Resource Management Review*, Vol. 11, 2001, pp. 299–326.

Meyer, J. P., N. J. Allen, and C. A. Smith, "Commitment to Organizations and Occupations—Extension and Test of a 3-Component Conceptualization," *Journal of Applied Psychology*, Vol. 78, No. 4, August 1993, pp. 538–551.

Meyer, J. P., and E. R. Maltin, "Employee Commitment and Well-Being: A Critical Review, Theoretical Framework and Research Agenda," *Journal of Vocational Behavior*, Vol. 77, No. 2, October 2010, pp. 323–337.

Meyer, J. P., D. J. Stanley, L. Herscovitch, and L. Topolnytsky, "Affective, Continuance, and Normative Commitment to the Organization: A Meta-Analysis of Antecedents, Correlates, and Consequences," *Journal of Vocational Behavior*, Vol. 61, No. 1, August 2002, pp. 20–52.

Miles, J., and M. Shevlin, M., "A Time and a Place for Incremental Fit Indices," *Personality and Individual Differences*, Vol. 42, 2007, pp. 869–874.

Murdoch, M., and P. G. McGovern, "Measuring Sexual Harassment: Development and Validation of the Sexual Harassment Inventory," *Violence and Victims*, Vol. 13, 1998, pp. 203–216.

Muthén, L. K., and B. O. Muthén, *Mplus User's Guide*, Los Angeles: Muthén & Muthén, 1998-2010.

O'Reilly, C., and J. Chatman, "Organizational Commitment and Psychological Attachment—the Effects of Compliance, Identification, and Internalization on Pro-Social Behavior," *Journal of Applied Psychology*, Vol. 71, No. 3, August 1986, pp. 492–499.

Ostroff, C., A. J. Kinicki, and M. M. Tamkins, "Organizational Culture and Climate," *Handbook of Psychology*, Vol. 12: *Industrial and Organizational Psychology*, Hoboken, N.J.: Wiley, 2003, pp. 565–593.

Pershing, J. L. "Gender Disparities in Enforcing the Honor Concept at the U.S. Naval Academy," *Armed Forces & Society*, Vol. 27, No. 3, 2001, pp. 419–442.

———, "Why Women Don't Report Sexual Harassment: A Case Study of an Elite Military Institution," *Gender Issues*, Vol. 21, No. 4, 2003, pp. 3–30.

Probst, T. M., T. L. Brubaker, and A. Barsotti, "Organizational Injury Rate Underreporting: The Moderating Effect of Organizational Safety Climate," *Journal of Applied Psychology*, Vol. 93, No. 5, 2008, pp. 1147–1154.

Rice, E. A., Jr., *AETC Commander's Report to the Secretary of the Air Force: Review of Major General Woodward's Commander Directed Investigation*, Air Education and Training Command, November 2, 2012.

Riketta, M., "Attitudinal Organizational Commitment and Job Performance: A Meta-Analysis," *Journal of Organizational Behavior*, Vol. 23, No. 3, May 2002, pp. 257–266.

Rock, L., *2012 Workplace and Gender Relations Survey of Active Duty Members*, Alexandria, Va.: Defense Manpower Data Center, Survey Note No. 2013-007, March 15, 2013. As of June 4, 2014:
http://www.sapr.mil/public/docs/research/2012_Workplace_and_Gender_Relations_Survey_of_Active_Duty_Members-Survey_Note_and_Briefing.pdf

Rock, L. M., R. N. Lipari, P. J. Cook, and A. D. Hale, *2010 Workplace and Gender Relations Survey of Active Duty Members: Overview Report on Sexual Assault*, Arlington, Va.: Defense Manpower Data Center, March 2011.

Rogers, K., and E.K. Kelloway, "Violence at Work: Personal and Organizational Outcomes," *Journal of Occupational Health Psychology*, Vol. 2, No. 1, 1997, p. 63.

Rosen, Leora N., Doris B. Durand, Paul D. Bliese, Ronald R. Halverson, Joseph M. Rothberg, and Nancy L. Harrison, "Cohesion and Readiness in Gender-Integrated Combat Service Support Units: The Impact of Acceptance of Women and Gender Ratio," *Armed Forces & Society*, Vol. 22, No. 4, Summer 1996, pp. 537–553.

Scarpate, Jerry C., and Mary Anne O'Neill, *Evaluation of Gender Integration at Recruit Training Command*, Brevard County, Fla.: Defense Equal Opportunity Management Institute, July 1992.

Schaubroeck, J. M., S. T. Hannah, B. J. Avolio, S. W. J. Kozlowski, R. G. Lord, L. K. Trevino, N. Dimotakis, and A. C. Peng, "Embedding Ethical Leadership Within and Across Organizational Levels," *Academy of Management Journal*, Vol. 55, No. 5, 2012, pp. 1053–1078.

Shane, J. M., "Organizational Stressors and Police Performance," *Journal of Criminal Justice*, Vol. 38, No. 4, July–August 2010, pp. 807–818.

Small, S., and D. Riley, "Toward a Multidimensional Assessment of Work Spillover into Family Life," *Journal of Marriage and the Family*, Vol. 52, 1990, pp. 51–61.

Solberg, M. E., and D. Olweus, "Prevalence Estimation of School Bullying with the Olweus Bully/Victim Questionnaire," *Aggressive Behavior*, Vol. 29, 2003, pp. 239–268.

Spector, P. E., S. Fox, L. M. Penney, K. Bruursema, A. Goh, and S. Kessler, "The Dimensionality of Counterproductivity: Are All Counterproductive Behaviors Created Equal?" *Journal of Vocational Behavior*, Vol. 68, No. 3, June 2006, pp. 446–460.

Spector, P. E., and Steve M. Jex, "Development of Four Self-Report Measures of Job Stressors and Strain: Interpersonal Conflict at Work Scale, Organizational Constraints Scale, Quantitative Workload Inventory, and Physical Symptoms Inventory," *Journal of Occupational Health Psychology*, Vol. 3, No. 4, 1998, pp. 356–367.

Stark, S., O. S. Chernyshenko, A. R. Lancaster, F. Drasgow, and L. F. Fitzgerald, "Toward Standardized Measurement of Sexual Harassment: Shortening the SEQ-DoD Using Item Response Theory," *Military Psychology*, Vol. 14, 2002, pp. 49–72.

Stephens, G. K., and S. M. Sommer, "The Measurement of Work to Family Conflict," *Educational and Psychological Measurement*, Vol. 56, 1996, pp. 475–486.

Stewart, S. M., M. N. Bing, H. K. Davison, D. J. Woehr, and M. D. McIntyre, "In the Eyes of the Beholder: A Non-Self-Report Measure of Workplace Deviance," *Journal of Applied Psychology*, Vol. 94, No. 1, January 2009, pp. 207–215.

Stockdale, M. S., M. Visio, and L. Batra, "The Sexual Harassment of Men: Evidence for a Broader Theory of Sexual Harassment and Sex Discrimination," *Psychology, Public Policy, and Law*, Vol. 5, No. 3, 1999, pp. 630–664.

Tepper, B. J., "Consequences of Abusive Supervision," *Academy of Management Journal*, Vol. 43, No. 2, April 2000, pp. 178–190.

Tett, R. P., and J. P. Meyer, "Job-Satisfaction, Organizational Commitment, Turnover Intention, and Turnover-Path Analyses Based on Meta-Analytic Findings," *Personnel Psychology*, Vol. 46, No. 2, Summer 1993, pp. 259–293.

Tjaden, P., and N. Thoennes, *Prevalence, Incidences, and Consequences of Violence Against Women: Findings from the National Violence Against Women Survey*, Washington, D.C.: National Institute of Justice and Centers for Disease Control and Prevention, November 1998.

Trevino, L. K., K. D. Butterfield, and D. L. McCabe, "The Ethical Context in Organizations: Influences on Employee Attitudes and Behaviors," *Business Ethics Quarterly*, Vol. 8, No. 3, 1998, pp. 447–476.

Ullman, S. E., M. M. Foynes, and S. S. S. Tang, "Benefits and Barriers to Disclosing Sexual Trauma: A Contextual Approach," *Journal of Trauma and Dissociation*, Vol. 11, No. 2, April 2010, pp. 127–133.

United States Code, *Uniform Code of Military Justice*, 2012 ed., Sup. 4, Title 10, Chap. 47, §801–946.

U.S. Department of Defense, *Report of the Federal Advisory Commission on Gender-Integrated Training and Related Issues to the Secretary of Defense*, Kassenbaum Report, December 16, 1997.

———. *Report of the Panel to Review Sexual Misconduct Allegations at the U.S. Air Force Academy*, September 2003.

———. *Report of the Defense Task Force on Sexual Harassment and Violence at the Military Service Academies*, June 2005.

———, *Report of the Defense Task Force on Sexual Assault in the Military Services*, December 2009.

———, *Department of Defense Annual Report on Sexual Assault in the Military: Fiscal Year 2012*, Vol. 1, Washington, D.C.: April 2013. As of June 4, 2014: http://www.sapr.mil/public/docs/reports/FY12_DoD_SAPRO_Annual_Report_on_Sexual_Assault-VOLUME_ONE.pdf

U.S. Department of Defense Directive 6495.01, *Sexual Assault Prevention and Response (SAPR) Program*, January 23, 2012, incorporating change 1, April 30, 2013.

U.S. General Accounting Office, *DOD Service Academies: More Actions Needed to Eliminate Sexual Harassment*, GAO/NSIAD 94-6, January 1994.

———, *Basic Training: The Services Are Using a Variety of Approaches to Gender Integration*, NSIAD-96-153, June 10, 1996.

———, *Military Housing: Costs of Separate Barracks for Male and Female Recruits in Basic Training*, GAO/NSIAD 99-75, March 1999.

Victor, B., and J. B. Cullen, "The Organizational Bases of Ethical Work Climates," *Administrative Science Quarterly*, Vol. 33, No. 1, 1988, pp. 101–125.

Williams, J. H., L. F. Fitzgerald, and F. Drasgow, "The Effects of Organizational Practices on Sexual Harassment and Individual Outcomes in the Military," *Military Psychology*, Vol. 2, No. 3, 1999, pp. 303–328.

Willness, C. R., P. Steel, and K. Lee, "A Meta-Analysis of the Antecedents and Consequences of Workplace Sexual Harassment," *Personnel Psychology*, Vol. 60, 2007, pp. 127–162.

Zohar, D., "Safety Climate: Conceptualization, Measurement, and Improvement," in B. Schneider and K. M. Barbera, eds., *The Oxford Handbook of Organizational Climate and Culture*, New York: Oxford University Press, 2014, pp. 317–334.